T0141697

Weathering the Storm: Sverre Petterssen, the D-Day Forecast, and the Rise of Modern Meteorology

Sverre Petterssen, 1944

Weathering the Storm: Sverre Petterssen, the D-Day Forecast, and the Rise of Modern Meteorology

By Sverre Petterssen

Edited by James Rodger Fleming

Historical Monograph Series
John S. Perry, Historical Monograph Editor
American Meteorological Society
Boston
2001

ISBN 1-878220-33-0
Library of Congress Control Number: 00-135856

Published by the American Meteorological Society
45 Beacon Street, Boston, MA 02108

Ronald D. McPherson, Executive Director
Keith L. Seitter, Deputy Executive Director
Kenneth Heideman, Director of Publications
Kate O'Halloran, Copy Editor
Leah Whalen, Copy Editor

Printed in the United States of America
by Sheridan Books

Contents

Illustrations

Foreword to the Norwegian Edition (1974)

It took me two full years to write this book. The manuscript, titled "Of Storms and Men" was initially written in English, and later translated to Norwegian [*Kuling fra nord: En vaervarslers erindringer* (Oslo: Aschehoug, 1974), or "Gale from the North: A weather forecaster's reminiscences"]. Teacher Jens Albert and his wife Karin Jensen have provided great help and support during the translation process. I am grateful to both of them.

My brother Anselm has, through his notes about life in *Nordland* [the northernmost region of Norway] in the old days and about the types of boats used, been of great help. His notes, together with notes about the Lofoten fisheries left behind by my father, have been indispensable during the process of writing the first chapter. My interest here has been the culture and living conditions, not so much the family's own affairs.

I am greatly indebted to Knut Knaus for the many illustrations he made especially for this book, as well as for his stimulating friendship. I would also like to thank Professor Tor Bergeron and Mr. C. K. M. Douglas for permission to simplify and reproduce their analysis of the storms portrayed in figures [2.7 and 19.1]. Likewise, I thank Professor J. Bjerknes for permission to publish family portraits of great interest. A special thank you goes out to publishers who have allowed me to use citations from books protected by copyright.

And above all, I would like to thank my many colleagues and students for their inspiration and friendship, as well as my wife Grace, who helped me so much in my work.

Sverre Petterssen
London, April 1974

Preface to the American Edition (2001)

Sverre Petterssen (1898–1974) was a giant in the field of weather analysis and forecasting and an international leader in meteorology during its formative era. In this lively and insightful autobiographical memoir, written just before his death, Petterssen shares intimate memories from his childhood in Norway, his education and service with the famous Bergen school of meteorology, and his extensive experiences in polar forecasting and as head of the meteorology department at MIT. The crisis of World War II comes alive in his passionate recounting of how forecasts were made for bombing raids and special operations. Notably, Petterssen was the only Norwegian-trained meteorologist involved in the contentious forecasts for D-Day. Petterssen's philosophy of life is also in evidence throughout the work, for example, in his friendship with the ill-fated writer Nordahl Grieg, and when he notes, "my personal problem was that of a pacifist–scientist in a technological culture where humanism has become little but an ornament."

After World War II, Petterssen served as head of the Norwegian Forecasting Service, scientific director for the U.S. Air Force Weather Service, and professor and chair of the department of meteorology (later geophysics) at the University of Chicago. Later in his career he served as president of the American Meteorological Society, member of the President's Science Advisory Committee, and chair of the National Academy of Sciences planning committee for the atmospheric sciences. Living in London in retirement in 1973, he voluntarily relinquished his U.S. citizenship to protest the Nixon–Watergate scandal.

Meteorology today is the beneficiary of Petterssen's fundamental work in weather analysis and forecasting—including his development of Bergen-school and upper-air techniques, and his discovery of what were later called "jet streams." Today's scientists are also the beneficiaries of many of the international programs in education, research, and service—such as the World Weather Watch—that are at the core of Petterssen's legacy.

Thanks are due to the following individuals and institutions for their valuable contributions to this project; they are listed in roughly chronological order. In 1998 Norm Phillips contacted me about his interest in Sverre Petterssen's autobiographical memoir and the difficulties he encountered in locating (and subsequently reading) a copy of *Kuling fra Nord*, published in Norwegian in 1974. He encouraged me to obtain a copy of Petterssen's original manuscript, written in English, and to bring it to the attention of the American Meteorological Society (AMS) for possible publication as a historical monograph. A subsequent note from Anders Persson put me in contact with the Petterssen family. I located the manuscript in the collections of the University of Oslo-Blindern. Librarian Grete Krogvold generously provided, free of charge, a complete copy of Petterssen's manuscript ("Of Storms and Men") and loaned me her copy of *Kuling Fra Nord* when none was readily available in the United States.

After reading the manuscript, and duly impressed by the quality of the story, I contacted Sverre's daughter, Eileen B. Petterssen, who, together with Sverre's widow, Grace Petterssen, authorized the project. Eileen's husband, attorney Anton Ørbeck, drafted the publication agreements. The family also graciously provided photographs for use in preparing the volume.

With the support and encouragement of AMS executive director Ron McPherson and of members of the AMS Committee on the History of Atmospheric Sciences, I commenced my editorial duties. My employer, Colby College, provided me with a sabbatical leave in 1999–2000, and I completed the project as a visiting scholar in History of Science at Harvard University. Kennie Lyman provided a safe haven and a quiet space in her home for editing. Thomas Sully's painting, *The Scholar*, peered over my shoulder as I worked. John Perry, historical monographs editor at AMS, read the entire manuscript. Colby student Grete Rød translated several pages of Norwegian text. Kate O'Halloran provided professional copy editing in a most timely and efficient manner. A special note of thanks is due to AMS deputy executive director Keith Seitter and his publications staff, who took care of innumerable details involved in producing this volume.

Small changes to the manuscript were necessary during the editing process. This was Petterssen's intention, as is evident from a two-page memo he prepared in January 1974 and attached to the original manuscript. Such changes include the following: (1) using American English;

(2) retaining all of the early chapters (which Petterssen thought might be "too Norwegian"); and (3) adding explanatory endnotes. In 1974 Petterssen wrote, "my preference would be to obtain the assistance of a professional editor who is familiar with pertinent American conditions and requirements." Since Petterssen engaged no such editor in 1974, the present editor has served in this capacity. The agreement between the AMS and the Petterssen family states, in part, "The manuscript shall be published in its entirety, with only those changes and annotations deemed necessary by the editor to enhance the meaning of the text."

Sverre Petterson's complete autobiographical memoir, published here for the first time in English, offers a fascinating view of a man, an era, and a science. It is my sincere belief that anyone interested in weather, World War II, the history of science, or Norwegian history will enjoy this book.

James Rodger Fleming
South China, Maine, June 2000

CHAPTER 1

Early Years

O dear little cabin, I've loved you so long,
and now I must bid you good-bye!
—Robert Service[1]

My earliest recollection goes back to the summer of 1900. I was then about thirty months old, and what I have always remembered so vividly is a single white daisy facing and almost touching a window in our tiny and crowded cottage. In Norwegian, a white daisy is called *prestekrave*—a priest's collar, or, more accurately, a priest's ruff.[2] What impressed me so much must have been the perfect whiteness of the petals and the symmetry of the flower. Later, when I was old enough for Mother (Fig. 1.1) to take me with her to church, I saw our kind old vicar in his ruff, and a strange sense of perfection and awe came over me. Soon afterward our vicar was succeeded by a tall angular man with a long dark beard that covered much of the ruff. He was not a nice man, I thought.

Through the *prestekrave* window we could see, on the other side of the narrow part of Eidsfjorden, the mighty mountain of Reka—a rugged peak rising high above the northeastward extension of the Lofoten chain of mountains, some 200 miles beyond the Arctic Circle. In these latitudes the daisies do not grow up to normal windowsill height, but our cabin was of a rather special kind, not uncommon in *Nordland* in those days but now extinct (Fig. 1.2). It was a low wooden structure reinforced by a turf wall that reached almost up to the windowsill. The wall was a double blessing: It kept out drafts, and it steadied the cottage when it was shaken by storms. The roof was made of rough boards and waterproofed by small, neatly laid sheets of birch bark, then covered by a deep layer of peaty turf. Much of the year snow added weight and protection against storm whirls and gusts. Mother, like most of our neighbors, usually kept a pot or two of geraniums in the window, and outside she had added a few squares of sod, with wild *prestekrave* in them, to serve as window boxes. On the whole, our little cottage was snug and easy to remember.

While I was too young to join my two elder sisters and two elder

Fig. 1.1: Mother, Maren Petronella Petterssen, née Hansen.

brothers in useful work, Mother used to take me with her on her many chores about our little farm. Sometimes she would make up poetry for me—just little rhymes, all about birds and animals, princesses and trolls. When I asked who had discovered poetry, she said it was Odin, who lived very long ago and had only one eye. A few years later I saw a man with only one eye; however, he had only lost one of his two eyes—the Odin idea was not nearly as dramatic as I had imagined.

Once during an early autumn storm, before snow had settled for the winter, our boat was carried on the wind and smashed near the cottage—some fifty yards from the shore. When I asked how storms are made, Mother told me that they all came from Reka, and I was quite certain that this was so.[3] On another occasion, a long-lasting storm wore down our cottage. When it had been rebuilt, expanded slightly, and supplied

Fig. 1.2: Dear little cabin: our small house close to Reka.

with two solid iron stays fastened to big rocks sunk in the ground, we all felt secure. The expansion and the stays may not have been considered signs of wealth, but somehow they pointed in the right direction.

It was about this time that I began to take a lively interest in my maternal grandfather who had moved in with us (Fig. 1.3). Though old, worn, and rheumatic, he was normally quite cheerful, but on winter evenings, when he felt "storm in the weather," he became rather remote and sometimes restless. Often he found solace in moaning; once in a while, he uttered words from sad hymns. On such occasions an aura of mystery seemed to surround him and I dared not ask him questions. It was customary in those days for old people to have their coffins made early; to die without was a disgrace—except for men who perished at sea. When Grandfather's coffin arrived my younger sister was well beyond the crawling stage and a new sister had recently claimed the family cradle. When I asked Grandfather "how babies happen," he began to talk about other things; when I wanted to know when he was going to use his coffin, he left me without a word.

Grandfather was too old and I too young to be useful, and he was quite glad to satisfy my curiosity about ordinary things. I soon discovered that he was fond of raw onion. Although such things were expensive, Mother let me give him a slice as a daily treat. In return, I learned a great deal about whales, porpoises, cod, and herring; the seasons of the Lofoten and the Finnmarken fisheries; the calendar of the migrat-

Fig. 1.3: Maternal grandfather, Petter Mikal Hansen.

ing birds; and the deeds of famous fishermen. I also heard of the big trolls that live inside mountains and steal princesses, and the homely little trolls that just poke fun at people. No one knew what northern light was really made of but on cold and clear winter nights, especially when the moon was new, the light became restless and we imagined we could hear its crisp cracking sounds; sometimes, according to legend, it might even flick down to earth and carry away a child. But this did not happen very often. On clear days, at the time when the sun is about to disappear for the winter, we often saw beautiful sunset arches near the horizon. Grandfather explained that God had put the rainbow in the sky for all people, but these arches were there for those of us who had to live without the sun through the dark and stormy winter.

When I asked Grandfather questions about himself he would twist the conversation on to other matters, and the question itself was soon forgotten. On such occasions he told me about my paternal grandfather, who had died very long ago. The family had lived somewhere in the Salten fjord not far from Bodø. Father's father was a very able man, a fisherman and a trader. He used to join in the Lofoten fisheries, buy salted or dried fish, sail his cargo to Bergen some 500 miles farther south, and return with needful wares. On January 25, 1870, he set out for Lofoten and was caught in a storm that lasted several days. With all reefs in[4] he drifted with the southeast gale across the outer reaches of the Vestfjord and, by luck, found Røst, the oceanward tip of the Lofoten chain of islands. The youngest member of the crew had frozen to death on the way, and the others had suffered serious injury. Grandfather Petterssen, being the headman, sat at the helm, and each overtaking breaker washed around him. As the boat rose on the next crest icy water ran out—also past the helm. The unwritten code of headmanship prevented him from handing over, and his kidneys suffered severe frost damage. Though he survived the journey, he never recovered his health, and died eighteen months later; Father (Fig. 1.4) was then a little more than one year old.

Soon the family fortunes dwindled and Father went to live with a nearby relative who was a kind man. At the age of seven, when his foster father died, he was sent north to some distant relative not very far from our "dear little cabin." Hard and toilsome years followed. At the age of thirteen he joined as a half-lot man [or shareholder] on a Lofoten team. Skill, energy, and an unbending competitiveness soon earned him promotion to a full-lot member. This enabled him, at the age of twenty, to buy four acres of land, a cow, and three sheep, and to build our cottage. A few years later, as a result of a lingering intestinal disorder, he had to give up fishing. The outlook for the rapidly growing family turned very bleak.

Neither Mother nor Father had had much schooling, but Father had a very valuable asset—a neat and artistic handwriting; he was good at figures and expressed himself clearly—orally as well as in writing; often neighbors came to Father for help with letters of consequence. Father took to trading, and became a traveling salesman. Although the business in this sparsely populated area was not very rewarding, it compensated more or less for the loss of a share as a fisherman who had no boat and tackle of his own. Grandfather was impressed with Father's skills and went out of his way to explain why Father had so little time to spend at home.

Fig. 1.4: Father, Edward Hildor Petterssen.

Though Grandfather was reluctant to talk about himself, there was much that could be gleaned from later conversations with his fishermen colleagues, their children, and grandchildren. Sometimes oral traditions last longer than written records. Even today Petter Hansen is remembered. Like many other lads of his time, he started as a half-lot man at the age of fourteen. Skill, luck, and untiring energy enabled him to obtain credit to buy boat and tackle and to become headman of a crew at an early age. Wear and tear and crippling rheumatism forced him to withdraw when he was sixty-nine. He had then plied the fishing banks for fifty-five years; of these he had been headman for forty-nine.

Grandfather was an astute observer; one of his talents was a keen sense of sounds. The Lofoten waters are infested with underwater reefs, and in stormy weather, the breakers are enormous and completely demanding. In snowstorms during the midwinter darkness there is not much to be seen, but Grandfather had the extraordinary ability to recognize the breaking reefs by their varying sounds. Old men said that Petter Hansen could navigate through snowstorms by his ears.

On one occasion, when Grandfather was fishing from an outpost about thirty miles from his home, a friend arrived with news. When Grandfather returned from the banks in the afternoon, he heard that his wife was in her second day of labor and that things looked critical. Grandfather, with his trusting crew, set out in an impending snowstorm and found their way through what they called "unclean waters," arriving at the doctor's residence at breakfast time next morning. Not unreasonably, the doctor refused to go, saying that it would be a madman's act to put to sea in such a storm. With his usual calm, Grandfather said, "I have enough men here to fetch you unless you choose to come with us. I also have a large sheepskin rug to wrap around you, and if you have good boots and dress well, you'll suffer no harm." The young doctor, far from being angry or frightened, asked his wife to cook these men a hot meal while he got ready. In spite of snow, storm, and reefs, Grandmother's life and the lives of her twins were saved. This episode was well spoken of, even long after my grandparents and Uncle Nicolai and Auntie Hanna were gone.

Not only within his own little community but also in a wider circle of Lofoten fishermen, Grandfather was a famous man. His fame rested entirely on his boatmanship and his skill and success in saving lives from storm-wrecked boats. At his funeral, an old and stooping fisherman was heard to say, "None could match Petter in stormy weather," and another old man answered, "Yes, he was the most daring and the most cautious of us all." Though a small, sparely built and lithe man, he knew how to command; his skill at obeying was probably never put to a real test.

But Grandfather was not so very different from his fellow fishermen. These men never thought of themselves as being particularly brave; skill, judgment, and *snarrådighet* [presence of mind, resourcefulness] were words they used to describe men of action. Life was hard; the men lived with danger; the forces of wind, waves, and breakers had to be faced; and even the best could be overpowered. God stood above all things and nature was His vicar. And God would help only those who did all they could not only to save themselves but also to help others.

The men fished in teams and shared the catch according to a time-honored formula in which the headman, the boat, and the tackle, as well as the men, were variables. The teams would compete among themselves and sometimes give vent to their differences, but when a storm struck, stronger bonds took hold and the sure way to disgrace was not to share in the danger. But the value of life was graded. Children, though lovable and loved, were plentiful and replaceable; women were more numerous than men, the storms saw to that. However, a breadwinner was something else, and the life of a widowed mother was a hard one. The women were trusting; if God saw fit to take a life at sea, He would also find some way of providing for the widow and children. And all knew that help came not from God alone; it came from God and hard work. In later years I realized that these religious beliefs were not relics of a distant past; they were perfectly tuned to the needs of this society. Nowhere have I seen such harmony among environment, objectives, and behavior.

When I was about eight, I heard Mother, in a conversation with a neighbor, refer to a storm as the one that left the community with six widows. This storm, which began on January 25, 1893, took seventeen lives out of the small parish of Øksnes. Mother had good reason to remember, for Father was in that storm, and it was only three days later that she heard that he was safe. Eighteen days after she received the message, her third child was born. Nowadays, we measure the force of the wind by instruments and express it in knots or some other impersonal unit. In 1804, when instruments were near-luxuries and terminology vague, Admiral Beaufort invented a scale of wind force in which 0 means *calm* and 12 means *that which no canvas can withstand.* What this learned admiral did was to use his own man-of-war as a wind-measuring instrument. The Beaufort scale soon found international acceptance. It was only after I had become a student of meteorology that I realized that Mother had her own scale of the force of the wind—a scale reflecting human suffering rather than the response of canvas.

Before the advent of the motor, the boats used in north Norway were essentially of the Viking ship type, widely known as the *nordlandsbåt*. It is a clenched-lap shell in which are fitted frames and ribs. Most of the strength derives from the overlapping clinkered strakes,[5] each shaped and twisted to produce the high stem and high stern profile of the Viking masterpieces. Internal frames and ribs are kept to a minimum to save weight, each clinkered overlap serving as a girder. The boat is strong and

Fig. 1.5: The keel is useful, even after a *nordlandsbåt* capsizes.

light; it can easily be hauled up on land when storms threaten; it performs well in shallow water; and it has strength to withstand the hammering of Atlantic storms and waves. The typical *nordlandsbåt* has only one mast and one large square sail that can be reefed [or taken in] from below to reduce the force exerted by the wind. Though the maneuverability is somewhat restricted, the controls are simple and well concentrated in the headman's hands. The size of the *nordlandsbåt* may vary a great deal. Each subtype has a name, but only the two largest need be mentioned here: the *fembøring* and the *åttring*. The difference in construction and shape between these variants is rather immaterial, except when it comes to the rescuing of storm-wrecked men.

The worst sea is normally encountered toward the rear of the storm, when the wind suddenly swings from a southwesterly to a northwesterly direction; the wind-driven waves then strike more or less at right angles to the swell, causing mountainous peaks and steep-walled breakers to occur. It is in such cases that boats most readily capsize. As the boat turns over, the men try to grab the ropes or the ledge and, eventually, crawl upward to get hold of the keel (Fig. 1.5). With sail and mast in the water the boat becomes quite stable, and it drifts in a direction right across the wind. The breakers then wash over it readily, and a firm grip on the keel is essential.

If the wind and waves are not too violent, the "easy" rescue operation is to sail in close to the capsized boat on its downwind side, throw a rope and haul a man to safety, and then repeat the operation until no one is left. In this kind of rescue, the *fembøring* and the *åttring* are about equally handy; the success depends entirely on skill in maneuvering a boat that has only one piece of canvas.

On the other hand, if the wind and waves are excessive, intricate maneuvering under the lee of the wreck becomes next to impossible, and only one possibility remains—a very risky one: to sail across the capsized boat and pick off the men. Normally two or more crossings are required to save a crew of four. In this kind of rescue operation, a *fembøring* is next to useless while an *åttring* may succeed if skillfully handled. The *fembøring*, being the heavier, sits deeper in the water and is, therefore, more of a menace to the capsized boat and its crew than is the *åttring*. Important also is the difference in the longitudinal profiles of the two boats. The shape of the forward part of the keel of the *åttring* is such that the impact is much lessened, as compared with that of the *fembøring*. It may be further reduced by approaching the capsized boat at a slant, and also by sailing over it near the stem or the stern, rather than over the midsection. The chances of success are greatly increased if, by luck or otherwise, the rescuing boat reaches the crest of a wave as it begins to break over the wreck. All this and many other things, including *snarrådighet*, skill, contempt of danger, and a deep sense of comradeship, go into the operation. No text can teach how it is to be done; the skill is handed down from father to son or from headman to crewman. And was there ever a fisherman who did not hope to work his way up to the helm? Circumstances may vary, but Caesar or half-lot man, all aspire to leadership.

In a graveyard near the place where Grandfather lived can be seen tombstones carrying the names of two brothers. Let their names be Hans and Lars. Hans was a merchant and outfitter and Lars, the younger brother, was headman of a boat owned by his merchant brother. Almost all fishermen were indebted to Hans for goods and tackle. They paid off their debts by selling their catch to him; residual debts were carried over from one Lofoten season to the next.

In January 1877, at the beginning of the season, a storm struck without warning and Grandfather was unlucky; there was no time to haul in the tackle, for which he was in debt. The storm continued, and on the third day, he realized all was lost. He then went to Hans to obtain extended credit, hoping that he and his young sons could make new nets and tackle. A contract was signed, involving his cattle as collateral and the selling of all his catch at a predetermined price until all debts had been cleared. The storm lasted five more days, and with new tackle work continued, but the contract proved to be hard to live with; the selling of all catch at the agreed price seemed to lead to slow ruin. Pride stood in the way of reopening discussion.

Some winters later, a violent storm struck—again without warning signs—as the boats were returning from the banks. This time Lars, sailing his brother's boat, capsized, and none of the passing boats found it possible to attempt a rescue of the crew. Then came Grandfather in his *åttring*, rode on a breaking crest and picked off one man, returned a few times and, eventually, all were safe. That evening, pride was no longer an obstacle; a few manly words settled all accounts. Grandfather, with his crew, is reputed to have saved seventeen men. Even if it should be true that such accounts tend to magnify the deeds, the number must have been considerable. On another occasion, when a boat capsized close to the shore, he, his son, and my mother (who was then fourteen) did the rescuing amid breaking reefs.

Only a few of the fishing outposts had sheltered harbors and the last evening chore was to decide whether the boats should be hauled up on land for the night. It was customary for the senior headman to decide for himself and for the others to take notice. Grandfather, when he had risen to seniority, introduced his own system. At the stroke of seven he was seen to walk onto a rock with a fine view in all directions. If, in his opinion, the weather was stable, he just returned to his shack. If, however, he deemed that there was a risk he would sing, in a loud and piercing voice, part of a funeral hymn, his favorite being:

> O grave where is thy victory?
> O death where is thy sting?[6]

And the men would haul up their boats. It was generally thought that Grandfather tended toward caution. On the other hand, it was appreciated that no boats had been damaged at anchor while he was senior headman. He was daring as well as cautious; and if his singing ability was no better than mine, his storm warnings must indeed have sounded fearsome.

In many respects, 1906 was a year of great change for our family. Father's interests had gradually shifted toward some small and developing places in the eastern part of Vesterålen. To be able to visit his home more often, he sold our four-acre farm in the stormy vicinity of Reka and bought a farm, almost twice as large, in the far more sheltered area of Vik. The farm was in a sad state of repair, so there was much work waiting for Mother and a flock of eager children. There were then five sisters and four brothers, ranging in age from seventeen to one.

Mother, who had considerable skill in the handling of the larger variety of *nordlandsbåts*, borrowed an *åttring* from a neighbor, collected her children, two cows, six sheep, and all movables, and set out for our new home. When the wind became uncomfortably strong and the cargo a bit unruly, she found shelter for the night, and arrived at Vik one day late. The family was again a going concern. In two months of midnight sun much can be accomplished if the priorities are right: hay for the cattle, potatoes for the people, peat and wood for winter fuel. Then the days began to shorten.

Vik had many advantages over our earlier home. The farm bordered a brook rich in small salmon and trout. The neighboring bogs and hills abounded in cloudberries, blueberries, and red whortleberries; the cost of sugar was the only limiting factor.

We lived about a hundred yards from the one and only shop within a wide area. Few were without credit, and the shopkeeper ruled with reasonable consideration over many families. Shopping, except for certain textiles, was predominantly a masculine business. The shop, with its telephone and various connections, was a source of news, useful information on catch and prices, and, above all, a trading place of gossip. I found the shop fascinating and offered to run errands for Mother as often as opportunity served.

Once in a while strange events were recounted, and some of them were well beyond my understanding. On one occasion one of our neighbors had returned from a visit to an outlying place and brought with him a tale that released salvoes of laughter. Oline was a fisherman's widow and her son, Ole, had not inherited the laudable qualities of his father. Ole was lazy and missed many a fine day when fishing was good. At the age of twenty, Ole insisted on getting married. His mother yielded, hoping that things would then change for the better. To marry in reasonable style, Ole felt he had to have leather shoes and a proper suit of clothing. Oline reluctantly borrowed a few kroner and effected the purchase. The morning after the blessed wedding turned out to be fine for fishing and, at six o'clock, Oline brought the good news to her son's attention. The response was minimal. After several gentle naggings, Oline decided to be aggressive and remind her son that they had borrowed money to buy the clothing. To this Ole responded, "Mother, sell the clothing; I shall never get up again!" Though all roared with laughter, I failed to see what was so funny. What a stupid thing to do, to sell one's clothing.

Schooling was no great problem. A political movement in the latter half of the previous century had resulted in a law making schooling, with seven years attendance, compulsory for all children. Though the counting of years was a simple matter, the term "attendance" required interpretation. In our sparsely populated area, there was only one schoolhouse—no bigger than the one to which Mary took her little lamb—and there was only one teacher. By dividing the schooling into three grades and allowing each grade two weeks on and four weeks off, one teacher could accomplish what the law demanded.

This practice had great advantages; there was always much work to be done about the farms, and children above the age of seven or eight were useful members of the household. Problems related to spare time, toys, and entertainment simply did not exist; we worked in teams, and each day provided something to be proud of. Work was more than a pastime: It was always matched by responsibility and sometimes rewarded by praise.

School commenced in the seventh year, and all respectable families (and there were none other) took pride in entering their children with a reasonable amount of skill and knowledge. In our case, teaching of reading and writing started in the fifth year. When entering school, we had to know the multiplication table up to ten by ten, the Ten Commandments, the Lord's Prayer, the Articles of Faith, and one or two of the shorter hymns. Since families were generally large, it was customary for the parents to teach the first and second child and to supervise the teaching of the third; thereafter some kind of chain reaction followed. One of my friends taught himself how to spell at the age of four by sitting across the table watching his elder brother at work. At the age of sixty, my friend (then in Buenos Aires) showed me that he could still read with fair fluency from any book held upside down.

In the normal course of events, school ended in the fifteenth year, not without a show of ceremony and authority. All school leavers had to appear at the vicarage for a four-week course in the doctrines of the

Lutheran Church and such moral and ethical extensions as church regulations and practices. At the end of the course, the vicar, in his ruff and gown, with parents present and with the children numbered and arranged according to their course performance, preached to his flock. The sermon summarized all lofty Christian ideals and gave guidance for the years to come.

Childhood ended here. The girls now began to wear long skirts and the boys long trousers. The farms were too small to support large families so, in most cases, school-leaving meant home-leaving also. The eldest son would normally start off as a half-lot man and later continue farming and fishing as his father had done before him. Some would seek employment in shops, offices, or budding industries in towns along the coast as far south as Bergen. Many became sailors, first on local ships and later, as age and skill increased, in steamers that plied the seven seas. In some respects, the expanding merchant navy compensated for the scarcity of tillable land. In the nineteenth century, and well into the first decade of the twentieth, men and women from *Nordland* emigrated in large numbers to the United States, most of them settling in the cold-weather region along the Canadian border. Five out of eight of my maternal aunts and uncles went that way.

My eldest brother, Anselm, left home at fifteen, worked in a shop in a nearby town and, later, acquired the schooling needed to make a career in the postal administration. When Anselm left, Petter was thirteen and I eleven. Most of the farm work fell to us. As a result, I felt promoted to a far more important position. I was no longer just an appendage; I had become an essential member of the household. I could talk about things to be done tomorrow and, without being childish, I could show that I was tired at the end of the day. Petter and I fished for home consumption, planted and lifted potatoes, cut grass, felled trees (small and scruffy birches they were), cut peat, and collected fuel for the winter. On the whole we managed almost as well as when Anselm was there.

When Petter was getting on toward fourteen, a sheep had to be slaughtered and mother agreed to let us do it. As usual, it was our ambition to do things as well as grown-up men. Not only had we seen slaughtering being done, we had carefully noted all details and rehearsed the operation before we went to work. The job was neatly done and the family beamed with pride.

Summer was always a hectic season. With poor soil, a small farm, and long and bitter winters, the number of cows and sheep had to be kept small. The balance between supply and demand, for cattle as well as for people, was often precarious, and the arithmetic one had to go through when the summer ended was often involved. Your best friends were in the barn; you greeted them with friendly words and you fed them as well as you could for, in turn, you and yours depended on what they produced. You petted each cow and talked nicely to her as you emptied her udders, and you gave little tidbits to the sheep for the wool that kept you warm. Mother would shear the sheep, clean and card the wool, spin the yarn, weave and shrink the cloth, and make our clothing. Our loom was busy about twelve weeks every autumn. The sheep were important creatures and so easy to look after.

If spring were tardy or if your late-summer estimates had been faulty, you wished you could explain to the animals why their hay rations were so meager. Boiled seaweed and fish heads, sprinkled with kibbled maize and stubble, were often served as a second course. Soon the snow would be gone. We loved our animals dearly, but our love did not degenerate into sloppy sentimentality. Petter and I did not feel badly about slaughtering the sheep; to us the difference between man and beast was plain. There was a time for children to leave home and a time for animals to depart.

While Anselm was thought of as a bookish sort of lad, Petter was exceptionally inventive in finding excuses for avoiding studies. The Commandments told us not to steal and kill, and not to commit certain other sins. Both Petter and I found these remote restrictions reasonable. However, custom demanded that we did not play or work out-of-doors on Sundays—at least not until mid-afternoon, when people had returned from church, and even then only sparingly. Sunday was a dull day except for those who liked to improve their minds. The reading aloud of religious articles was a common activity in the neighborhood. Many of the stories were of the type that described the virtues of a teenage girl helping her sick mother, or the good deeds of a hard-working son of poor parents. Mother neither encouraged nor discouraged such enlightenment, and Sunday remained a problem.

It was here that Petter's inventiveness came in handy. Where we lived, going to church meant rowing or sailing. On fine Sundays in spring and summer, Petter and I set out in the family *fering*[7] at the appropriate time and spent hours on nearby islands, sometimes catching a fish or two on the way. On many of these excursions we gathered down and feathers and collected eggs, marveling at their varying shape and color. The art of nest-building did not escape our interest, and once in a while we discovered that some eggs have an oilier taste than others. The threatening shrieks and dives of the parent birds just added to the excitement.

Grandfather had revealed an almost mystical feeling for birds and their behavior in relation to the weather, and his account of the calendar of migrating birds now came fully alive. The first to arrive is the black guillemot (or tystie) who comes in late February, when the days are lengthening most rapidly. We were always anxious to report the arrival of this first visitor. Though not very large, it is easily recognized at a distance by the oval patch of white on its wings and its high-pitched call. In early March, the noisy kittiwake appear in great numbers, but as they look just like small seagulls, little attention was paid to them. Real excitement attended the arrival of the kjeld (or oystercatcher), who is a swift flyer and was believed to have come all the way from Egypt. With its long red beak, red legs, and black and white attire, the kjeld was easy to recognize and very interesting to watch. Old people thought of the kjeld as a sacred bird, though none could explain its attributes except that it seemed to be able to predict the arrival of different kinds of storms. When it flies about in a worried and irregular fashion and sends out shrieking calls, a storm from the Atlantic is not far away. On the other hand, when it issues the same sounds while running restlessly about on the shore, a storm from the Arctic is in the offing. Often the kjeld just sits or walks about quietly, or sings with a lark-like voice; this may be a sign of stable weather—or perhaps the bird is just relaxing after an irksome forecasting shift.

The really busy period is the week following the spring equinox. In rapid succession come the razor-billed auk, the common guillemot, the precious eiderduck, the disreputable cormorant, and the strong-billed puffin, who defends its nest with great skill and cunning. When the puffin arrives the days are a little longer than the nights but spring is still far off. The greylag goose [*Anser anser*] comes without much ceremony in the middle of April, when an arctic variety of spring is in the air. But real spring does not come until the first week of May, when the arctic skua is well established. The buds of the rowan and the birch are then swelling rapidly, and even the stunted juniper shows signs of a new cycle. The last bird to arrive is the common tern; it waits until the midnight sun has ceased to be marginal. By definition, summer begins with June.

Many of the fish-eating birds do not migrate and some of them are rarely seen on land. The havhest (or fulmar), which is thought of as a gull but is really a relative of the albatross, has its summer home in Iceland, though some nest on a few of the outermost Lofoten islands. Once in a while a havhest is carried inland on storm winds and may then be found in a state of exhaustion. Most fishermen thought of the havhest as a bad omen and called it the *stormfugl* (storm bird). Another rare bird is the huge white-tailed eagle, which lives on fish and sometimes coughs up neat balls of scales. Its nests are built of branches and twigs and snugly lined with seaweed and moss. Petter and I nourished hopes of robbing a nest before the young ones could fly, but our ambition never came true.

The migrating birds that come to *Nordland* leave early; by the end of August few are left. Among the last to leave is the greylag goose, which may remain until late September; autumn storms, snow, and long nights are then not far off. Real darkness is heralded by the eagle; as snow accumulates inland some go south, while others just move toward the coast, to mountains of the Reka type, where the wind sweeps snow away and access to fish is easier. But some of the birds that ought to migrate do not do so. Weak yearlings stay behind the first winter and seem to manage quite well. Elderly eagles and other old birds may decide to remain come what may.

A few years after our move to Vik, Father began to travel for a wholesale firm in Bergen. On his annual visits to his employer he brought home wares of different kinds: pencils, pens, fancy writing paper and lined envelopes, scented soap, and other things that could easily be carried about. Petter and I, being in different grades, had two weeks out of six off from school at the same time. In autumn and winter, when work on the farm was not too demanding while cash, as usual, was in short supply, we often set out on peddling expeditions to outlying areas, usually a short day's rowing from home. We both enjoyed these excursions, and as our backs grew stronger, Father added religious books to the load. The demand for such books proved considerable. Father, now about forty, was what may be called a traveled man. He knew and used freely a number of foreign words that had crept into the Norwegian language and were often used as a status symbol. Indeed, a *fremmedordbok* (dictionary of foreign words) was amongst the few books we owned. Father explained that a person who sells religious books is called a *kolportør*. I became keener than ever to go on these hawking expeditions.

Once in a while I tried to read the books we were selling but found little clarity in any of them. Petter, being far more cynical, just scoffed at my misgivings and said such books were written for old people. Though I pretended concurrence, his explanation did not quite satisfy me. In one house where we stayed overnight I came across an old, worn, and beautifully illustrated translation of a book attributed to Thomas à Kempis (*Imitatio Christi*). The few pages that I read that night made a deep impression on me. These stirrings never led to any major resolve on my part; they were significant not so much for their intensity as for their unwillingness to leave me in peace.

Peddling was hard work, but it was also stimulating: new places, different people, a variety of responses and, above all, an unvarying hospitality. Life about the farm was generally uneventful and even the arrival of a couple of boy peddlers would break the monotony. Petter and I soon found that there is a time for this and a time for that. Fancy writing paper and religious books sold best during the Lofoten fishing season, the books being bought almost exclusively by married women, and the paper by hopeful girls. The demand for scented soap (locally called *godluktsåpe*, literally, "good smelling soap") was more steady, though sales seemed to increase in anticipation of the return of the menfolk; here, wives were as eager as unmarried girls.

Petter, who was no lover of schooling, left home when he was barely fourteen. His first job was that of a junior deckhand on one of the small steamers that served the Lofoten–Vesterålen area. Every second week his ship stopped for a few minutes at Vik. But Petter was in a hurry—he could not wait to grow up. Soon he was on a steamer that plied the North Sea. From frequent postcards we learned much about the fine buildings and monuments in London, Amsterdam, Hamburg, and other places. After a year or two, occasional letters and pictures that came from San Francisco, Hong Kong, Melbourne, and other faraway places told their story; the intervals between the messages increased with the years.

With or without news of Petter, Mother was often uneasy. At the age of three he had fallen backward into a pan of scalding water and was barely saved from serious injury. At the age of seven he fell through thin ice and was rescued in the nick of time—mostly through his own *snarrådighet*. Petter was probably as close to Mother's heart as any of her children. He was now so far away—and he was so young when he left.

Petter's departure, as well as time and circumstances, led to many changes. My eldest brother Anselm was already on his own; my youngest brother was only five, and Father was always traveling. On the peddling expeditions and on occasional visits to church and to some small nearby trading centers I had seen and heard educated men: the vicar, the doctor, the apothecary, the judge, and others. Even the schoolteacher carried with him an aura of learning. I had also seen what I then considered to be wealth and poverty: The merchants were wealthy, and the fishermen generally poor. The wealth was certainly not excessive—it was well within the range of my imagination. The poverty I knew best: We had much of it at home, but it was not ugly poverty, it was entirely of an economical kind; real human poverty, with or without wealth, remained for me to discover.

In my thirteenth year I began to think of my own future. It was not for me to decide and tell Father and Mother that this or that was what I had chosen—but the urge was strong and my dreams lively. I began to think in terms of two kinds of men: the educated and the uneducated. Poverty and wealth were unimportant; all things good and lofty seemed to spring from learning. One day when I saw Petter operating the noisy winch of the smelly little ship in which he was working, I felt that he was surrounded by ugliness. Then and there, I decided not to follow Petter; I wanted to do all I could to become an educated man. It was not that I wished to become a lawyer, or a doctor, or a clergyman, or a scholar: I had little understanding of the root characteristics of these fields of learning. It was the professional class as a whole and the lives its members led that appealed to me very strongly indeed. In retrospect, it appears that my resolve to seek education was among the firmest decisions I ever made.

There were other stirrings. Some of the impulses came from the outer world; even the firmament was lively. In May 1910 the mighty Halley's comet crossed the sun, and its broad tail swept almost right across the sky. No one knew when the earth would get into its poisonous fumes, but to many the signs were unmistakable. As the comet withdrew and the year came to a peaceful end, rumors told of other comets, entirely new creations, that had been seen in various places, especially in the Southern Hemisphere. No one knew how it would all end. However, the vicar maintained that comets were harmless things, created by God; all learned men supported his view.

The men at the boathouses talked about the war in the Far East. War was considered to be manly and heroic. Admiral Togo was a great hero: He had defeated the Russians! During a century of union between Norway and Sweden, the traditional Swedish fear of their neighbor to the east had spread into *Nordland*, even to the extent that visiting strangers were suspected of spying for the czar. The noises created by Kaiser Wilhelm reverberated far beyond the Arctic Circle: The Germans were very clever people, all agreed; Father was all for them. Uncles and aunts who had settled in northern Wisconsin sent home encouraging news. Uncle Nicolai had gone to Alaska and rumors had it that he found sizable amounts of gold—real gold! Soon all of them would move west to the far more prosperous state of Washington. An awareness of the world around us began to take hold of me.

> To give space for wandering is it
> That the world was made so wide[8]

In 1905 the union between Norway and Sweden had come to an end. The national upsurge, which began early in the nineteenth century and gathered momentum after about 1850, was the root cause of the freedom that people now enjoyed so much. The leaders of this movement were men of learning—the upsurge was carried on cultural rather than political wings. The heroes in Norway were not politicians, admirals, and generals: They were writers like Henrik Ibsen, Bjørnstjerne, Jonas Lie, and Alexander Kielland; musicians like Edvard Grieg and Ole Bull; professors in the humanities and sciences.

In earlier centuries *Nordland* was very much an appendage, a stepchild poorly spoken for in the councils of state. After the separation from Sweden, the national upsurge took on a regional garb: *Nordland* was to be an equal partner with the other provinces in the business of Norway. Schools were to be improved, postal and telegraph services were to be expanded, and even the idea of a trunk railway connecting *Nordland* with Oslo began to receive serious consideration. These new ideas were more than political stunts. The vicar, the doctor, the judge, in short, the educated class, led with knowledge and wisdom. The fishermen discussed the problems of the future while mending their tackle.

The merchants (though some may have looked on progress with the eye of a Swiss innkeeper for beautiful scenery) were keen; in their shops the spittoons filled while the men vented their views. All seemed to agree that the way to "equality with the South" was through an improved school system: more secondary schools, high schools, and other schools right in *Nordland*. Though these mild breezes did not stand out to me as representing a coherent trend, something new, a kind of spring, seemed to be in the air. I was probably much influenced by what I heard.

Once more the family circumstances changed. When Petter left home, Father began to think of selling our little farm at Vik and moving the family to the south; Trondheim became his choice. Better schools and more opportunities beckoned; perhaps, one day, he might stop traveling and develop a business of his own.

The family left in June 1911, but I had to stay behind to work for the summer as a galley boy on the local steamer where Petter had started a year earlier. I arrived in Trondheim three weeks after school had started and found myself, rather unexpectedly, to have become a problem child. The headmaster did not know quite what to do with me—a boy sailor of thirteen, coming from remote *Nordland* where schools had only three grades, to be placed in a modern school with seven grades and school six days a week, every week. He decided on the fifth grade, but a few days later I was quietly moved up one step. My classmates were curious. I was not immediately accepted; my clothes smelled of fish, I was told. After one or two fights with bigger boys, good relationships developed.

Trondheim proved a disappointment. The family finances remained a problem. Here we could not go to the river or the fjord to catch fish, nor could we go to the cellar to fetch potatoes; even the milk had to be bought. My summer's earnings came in handy, as did small profits from selling newspapers in the streets. But newspapers are notoriously unreliable: On rainy days there are fewer customers and the papers get soggy; once in a while you lose rather than gain.

After a few weeks, and with autumn in the air, I left the newspaper trade and took a half-time job as an errand boy in a grocery shop. School from eight to two and work from two-thirty to eight filled the day well. Every Saturday I received my pay: fifty cents. Five months later I transferred to a china shop where I had better pay, fewer deliveries, less weight to carry, and the absorbing work of packing breakable things. In later years, when my work took me to many lands, I never trusted my glass and china to commercial packers—even on occasions when some employer defrayed all expenses.

The law, which required seven years of schooling, had a clause that permitted departure at the age of fifteen, regardless of attendance. Taking advantage of this clause, I left school four months before finals and took a full-time job in the telegraph office earning eleven dollars a month, with a small allowance for my uniform.

Although these activities did not add up to support for my educational schemes, they served a useful purpose. Even before I left school I had developed a plan which, more or less, took me to my goal. Right in Trondheim was a school for noncommissioned army officers: entrance at seventeen, everything free, three cents pay a day, duration three years and eight months. In addition to customary military subjects, the program included a regular secondary school education. There were eight such training centers in Norway, with a total annual output of about 300 students. Four scholarships, each 125 dollars a year, were awarded to encourage students to continue through high school (*gymnasium*) to qualify for entrance to the War College.

Though the chance of getting one of these scholarships was no better than one in seventy-five, there was never any doubt in my mind. I went on to the *gymnasium*; the scholarship and a small pay as a sergeant in the reserve took care of all necessities; and a thoughtful headmaster allowed me to work on some kind of a double-shift arrangement to save one year. I was now only three years behind normal schedule.

During the *gymnasium* years my feelings about a military career changed. Under the influence of one of my teachers—a German renegade priest who had broken with the Roman Catholic Church and married a nun—I became deeply interested in studies in the humanities. The galumphings of the wartime leaders had become rather distasteful, and what I then considered to be a set of real values had taken shape. Literature had become a dominating interest: Goethe, Heine, Lenau, Schiller, Dickens (which I read in German translation), and Shelley (which I tried to read in English) took most of my spare time. There was so much to look forward to. The two busy high school years had opened a gate to a new world of riches. Unfortunately, perhaps, the gate was not for me to enter.

Though a military life had become distasteful, the obligations attached to the scholarship had to be met somehow. The war was followed by Wilson's peace plan, the establishment of the League of Nations, general talk of disarmament, peace forever, and so on. In this avalanche of sham and slogans it was difficult for me, and perhaps also for others, to distinguish between noble intentions and camouflaged expedience.[9] Whatever the truth might have been, the fact is that the postwar problems did not stir me very deeply. I was too occupied with personal problems.

A high-ranking army officer who had developed a real interest in my future advised me very strongly to try to escape from a military career and to study at Oslo University. He helped me write an application to the Ministry of Defense to be relieved of my obligation to enter the War College. My benefactor looked at the problem from a practical rather than a philosophical point of view: In the military stagnation was bound to come; I could never hope to have the satisfaction of being an expert in military affairs; and most likely I would reach retirement age without attaining high rank.

The application was successful: I was relieved of my obligation to go through the War College but, instead, I had to serve six extra months as a drill sergeant—with pay. I was immensely pleased with the outcome and grateful for the help I had received.

The Mount Pisgah from which I thought I had seen the Promised Land turned out to be a mirage. Although tuition at Oslo University was free, living expenses were high and I could meet them only by earnings from part-time work. While it was possible for science students to find odd teaching jobs, work for students in other fields was extremely difficult to find. Being faced with a Hobson's choice,[10] I had no great difficulty in adjusting my desires.

CHAPTER 2

Apprenticeship

Imagination is more important than knowledge.
—Albert Einstein, *Saturday Evening Post, 1929*

Meteorology, as a science and a service, was not sufficiently developed to play any important part in the decision making of the First World War. However, as the war was nearing its end, a flash of light from Norway attracted the attention of meteorological circles. Although the initial response was skeptical rather than enthusiastic, the new ideas—and they were ideas rather than firm building blocks of knowledge—have stood up to the scrutiny of science and the test of time. Indeed, many of the household words that now appear in television and radio weather summaries have developed from the ideas that began to emerge in 1918.

The center of activity was a small group of scientists, unmindful of conventional procedures but eager to explore new approaches; the locale was Bergen, Norway; and the leader of the group was no less a person than Vilhelm Bjerknes (Fig. 2.1).[11] Though the roots of his work may be traced back to his father and major scientific developments in the previous century, the first visible spark with a direct bearing on weather forecasting came from his young son, Jacob (Jack), then barely twenty (Fig. 2.2).

Other early members of the Bjerknes group were the mathematically inclined Halvor Solberg (Fig. 2.3), who liked to foreshadow elaborate solutions to a great variety of intricate problems, and the meticulous physicist, Tor Bergeron (Fig. 2.4), who tried to explain the total complexity of weather systems rather than the workings of their individual components. On and off, the group was agitated and enriched by stray visits of the young and restless Carl-Gustaf Rossby (Fig. 2.5), who usually left behind an atmosphere filled with soap bubbles—a remarkable number of which never burst. After about 1925, long and productive working visits by Erik Palmén (Fig. 2.6) added new knowledge and left seeds that grew, sometimes slowly, into long-term trends. From time to time, visitors from many lands came to Bergen; some were attracted by the theoretical works of Vilhelm Bjerknes, others, with more direct

Fig. 2.1: Vilhelm Bjerknes (1862–1951), theoretical physicist, explored the laws for movement in air and sea, and laid the foundation for modern forecasting.

Fig. 2.2: Jacob (Jack) Bjerknes (1897–1975), meteorologist, advanced his father's work, discovered the polar front, and spearheaded the development of the so-called Bergen methods in forecasting.

interests in current meteorological problems, sought the company of Jack Bjerknes, Tor Bergeron, and the forecasters in the weather center.

By the mid-twenties, what had commenced in 1917 as a small Norwegian compact had developed into an incohesive Scandinavian group. Palmén's family roots and academic home were in Finland, and he chose to remain there. Bergeron returned to Sweden. Rossby left to develop the first academic department of meteorology in the United States at the Massachusetts Institute of Technology (MIT). Much later, in 1939, Jack Bjerknes went to the University of California at Los Angeles (UCLA).

And the great leader of the group? In 1926, at the age of sixty-four, Vilhelm Bjerknes chose to transfer his activity to Oslo University, where he could more readily gather around himself advanced students with interests centered on his lifelong work: the theories that govern the physical and mechanical processes in the atmosphere and the oceans. Success always attended what Vilhelm Bjerknes set his mind to.

Fig. 2.3: Halvor Solberg (1895–1974), one of Vilhelm Bjerknes' original research assistants, who helped develop the polar front theory of cyclones.

Fig. 2.4: Tor Bergeron (1891–1977), musician, humorist, and master analyst of the Bergen school.

Fig. 2.5: Carl-Gustaf Rossby (1898–1957), multifaceted researcher and inspiring teacher, whose work greatly influenced forecasting aided by electronic computers.

Fig. 2.6: Erik Palmén (1898–1985), bon vivant, oceanographer, meteorologist, and untiring researcher of the upper layers of the air.

The first visible spark that had focused attention on the Bergen group in 1918 was Jack Bjerknes's discovery of the internal structure of the large migratory weather systems that meteorologists call cyclones. Jack found that the distribution of weather, wind, and temperature through a typical cyclone is discontinuous rather than continuous, and the discontinuity, or zone of transition, is an important part of the physical mechanism that keeps the cyclone going.

Jack proposed a blueprint, or a model cyclone, that went beyond the mere "anatomy" of storms and described in broad outline their "physiology." In addition to the contrast in temperature between the adjacent air masses, there is a corresponding contrast in air density, which, in turn, means a contrast in potential energy that may be used to create kinetic energy—the all-important storm winds. Jack's model indicated, though in a somewhat crude manner, how cloud systems, with rain or snow, are maintained, and how the movement of the storm as a whole might be related to the air currents in the warmer of the two air masses.

No one could have been more pleased than Jack's father, for the model pointed toward applications of Vilhelm Bjerknes's famous (but sparingly used) circulation theorem, developed when Jack was in his cradle. Equally pleased might have been Professor Max Margules (in Vienna), who was the first to consider the possible energy sources of the winds. But, alas, due to malnutrition caused by the hardships of the First World War, Margules's health was then rapidly declining. He died in 1920.

In retrospect it may well be said that Jack's model was a remarkable contribution, in part because of the volume of reliable knowledge that it conveyed, but more so because it represented a daring and imaginative approach to the understanding of an important and exceedingly complex meteorological phenomenon.

Extensions and improvements were soon to follow. In the early twenties Jack went on to produce, with Halvor Solberg, a set of evolutionary models showing the characteristic features of cyclones. The storms had a typical life history: They would start from meager beginnings, develop to peak intensity, and then dissipate their energy and lose their identity.

At about the same time, Bergeron developed his theory of how temperature contrasts form, how typical air masses develop, and how millions of minute cloud droplets join to form a typical raindrop, or, in meteorological language, how rain becomes released from clouds. In a manner of speaking, the cyclones were thought of as battles between polar and tropical air masses; the line of separation between these masses was traced on the weather charts. Naturally, in the lingo of the First World War, the line of separation was called the front, or more specifically, the Polar Front.

These were exciting years—years of growth, and also years of speculative creations. The processes and phenomena of cyclones were treated in all three dimensions, but upper-air observations were sadly lacking. In the absence of such observations, there was room not only for hypothetical thinking but also for skepticism and, on occasion, destructive criticism. Expressions like "What is true is not new," and "What is new is not true" were not altogether untypical of the response from some quarters, especially from leading German-speaking meteorologists. Also, perhaps, a slight streak of smugness in some of the members of the Bergen group tended to stimulate opposition.

Toward the end of my second year at Oslo University my geography professor surprised me by inquiring about my ambitions and plans; he hoped I would choose to major in a geophysical subject, and recommended meteorology as a field with a bright future. In the normal course of events the choice of a specialty would not be due until one or two years later; I was really not prepared to indicate a preference. Moreover, I had not considered meteorology—not even remotely—as a preferred field of study. The university had no chair in the subject; the work in progress in the Bjerknes group had not trickled down to the undergraduate level; and the texts on meteorology were old and sadly outmoded—full of tabulated data and dreary descriptions of individual phenomena, with hardly any reference to the laws of physics. Examples: "Rain falls in drops, never as threads"; "The raindrops are solid," in the sense that they contain no internal cavity; "Bloodrain is rain of a reddish color, caused by red dust whirled up by strong and hot winds over the Sahara desert"; "A cyclone is normally divided into four quadrants . . . ," and so on. These were things not to understand but to memorize. Meteorology was taught as a branch of physical geography which, at Oslo, included general geography as well as a smattering of geophysical subjects ranging from the earth's core to the northern-light region and near-space. It was a hopeless subject to teach, even for our brilliant and hardworking professor.

Though my response reflected no enthusiasm for meteorology, the professor persevered and explained what was in progress in Bjerknes's group. An international seminar of four weeks' duration was planned in late spring 1923. Bjerknes was anxious to attract "young talent" to meteorology, and he was prepared to provide some financial support for one or two suitable students who might wish to attend the seminar. Two weeks later Tor Bergeron came to Oslo, and, after a stern interview, I was selected. Undoubtedly, not only the program but also the financial support influenced my decision. The last few months of an academic year were always a lean time.

Finding the seminar both stimulating and disappointing, I saw no easy way to come to a decision about my own future. Vilhelm Bjerknes was recuperating at a German spa, and Jack was working in Switzerland—trying to spy on the behavior of the upper air by analyzing observations from mountain observatories. Neither would return before the seminar came to an end. In two earlier years the seminars had been well attended, and a major conference of the International Commission for the Exploration of the Upper Atmosphere in 1921 had been an enormous success. However, the present seminar (in 1923) had not attracted many participants from foreign countries; with one notable exception, the foreign participants were meteorologists mainly in an administrative sense. The exception was Carl-Gustaf Rossby, who lectured on some mathematical subject related to solar radiation. His contributions were nowhere near the central theme of the seminar, and I followed his lectures with great difficulty. Our long and lasting friendship began to develop much later.

Bergeron headed the seminar, and his lectures on weather analysis and fronts were both stimulating and enjoyable. Bergeron was a scientist and an artist. There was a human touch to everything that he said or did. Here was something new—in substance as well as in form—something really attractive. There were other contributors, but few of them rose much above the current textbook level. Their frequent use of phrases like "Experience shows . . ." and "For example . . ." proved nothing, and did not always sound convincing.

In 1923, there were no upper-air observations, and reports from ships on the Atlantic were sparse. As a result, there was much leeway in the analyses and, therefore, much to be discussed. On the other hand, Bergeron's skill in interpreting widely scattered observations and combining them into coherent analyses was impressive. His astuteness in inferring vertical structures from sea level analyses was simply uncan-

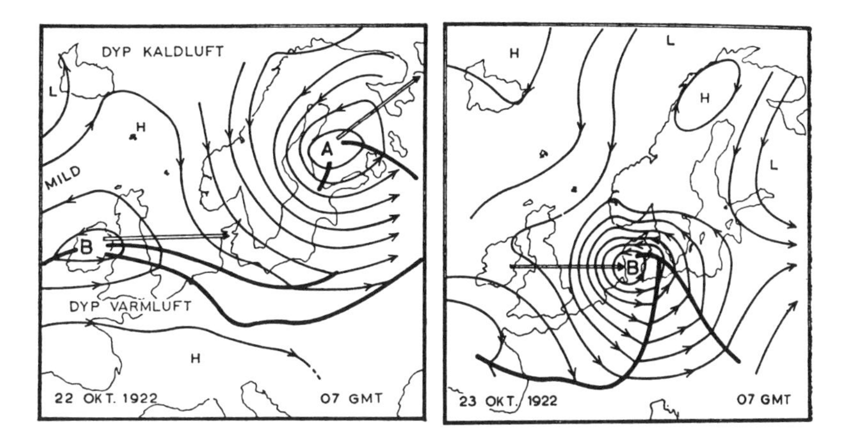

Fig. 2.7: The Ryder Storm of October 22–23, 1922. The sudden development of storm B was well predicted by Tor Bergeron. It was his account of the storm that inspired Petterssen to pursue a career in forecasting.

ny—something, I thought, I could hardly hope to be able to match. Science is one thing, and what we often—and loosely—call art, is something else. In Bergeron's presentations, the border between the two was sometimes diffuse.

While my geography professor had expressed himself in terms of weather forecasting, the overarching theme of the seminar was really weather analysis. To explore and understand a phenomenon was science; to analyze data and discover a phenomenon was science; but to predict the future growth and movement of a phenomenon was an art—an acquired skill, something that would come with experience.

As the seminar was nearing its end, an alternating current of pros and cons—shall, shall not—took hold of me. In the end I decided to call on Bergeron, tell him of my doubts, and indicate that I might not choose meteorology. As I entered his office I found him humming a merry tune and almost gloating over some weather charts—a lovely storm, now seven months old. Before I knew it I had become absorbed in his fascinating tale of the Ryder Storm—so named after a meteorologist who had failed to predict the storm; certainly not a diplomatic choice of name.

On October 22, 1922, a major storm A (Fig. 2.7) had crossed southern Scandinavia, entered Finland, and begun to weaken. At the same time, a minor disturbance B had formed over Ireland. Bergeron, then working in Stockholm, thought he could "smell a rat"; to him, storm B, with its strong temperature contrasts and rapidly failing pressures, was a potential giant. He issued warnings of gale winds, which he later strengthened to full storm. Bergeron's warnings from Sweden and those issued in Norway were soon and generously recognized as a spectacular success of the methods developed by the young and aspiring Bjerknes School of Meteorology. This recognition was magnified by the failure of meteorologists on the continent to predict the storm. Worst off was a coastal city where storm signals, lingering too long after the passage of storm A, were lowered just before storm B struck with devastating force.

Though fascinating in the extreme, Bergeron's account could not be quantified—expressed in numbers. We had before us all the data, the temperatures, winds, pressures, and so on, but we could not compute what we wanted to know: the rate at which storm B would intensify and the direction and speed of its movement. We couldn't, and no one else could; here was a blank page in the book of science.

Though weather forecasting may be said to be a science and an art, we may well ask where the former ends and the latter takes over; almost always conceptual borders are diffuse. We need not here be concerned with such exalted definitions of art as those proposed by Joseph Conrad, Henry James, and others;[12] but some wisdom may be gleaned from the writings of Lord Kelvin:

> When you can measure what you are speaking about and express it in numbers, you know something about it; but when you cannot measure it, when you cannot express it in numbers, your knowledge is of a meager, unsatisfactory kind.

While listening to Bergeron's account of the storm, thoughts like these flashed through my mind. Surely the concept of art (whether exalted or plain) should not be used as a garb of elegance to cover our ignorance of nature. Kelvin was right: The role of science is to shrink the blank areas in our knowledge of the physical universe. The word "art" and other words with a double meaning tend to encourage muddled thinking. My grandfather, who was said to be able "to navigate through snowstorms with his ears," practiced an art, but scientists must strive to quantify. On the spur of the moment, I decided to become a meteorologist, and with a considerable degree of single-mindedness, I decided to concentrate on the problem of prediction, predictability, and related areas. It was, perhaps, fortunate that I did not then realize the magnitude of the problem. But as the years passed, I learned to be satisfied with progress in homeopathic [minute] measures.

Financial support was now made available for me whenever I had time to work with the Bergen forecasters. By careful planning and management I could save enough to accommodate lengthy visits to Oslo University. No longer did I have to hunt for odd teaching jobs; I could even resign from the army. I know that "*impletus venter non vult studere libenter*" [an overfilled belly will not study willingly]. After the seminar came fourteen good months. I felt free, as I never had before.

My first meeting with the Bjerknes family was far less awesome than I had expected. I concluded that I had been invited to be looked over, and since I had but little experience in social affairs I felt awkward sitting next to the professor at table. One of the sons was studying to become an architect, and le Corbusier's early ideas were already making their influence felt in well-informed circles in Bergen. The conversation naturally turned to dwellings—old and new. In Norway, where the climate is rather severe and almost all fuels have to be imported, conservation of heat had long been an important consideration. With effective double-glazing and tight-fitting doors the air tends to get stale and the rooms stuffy. The young architect was all for large windows and good ventilation. Then, suddenly, and in the tone of an examiner, the professor turned to me and asked, "Why is the air so pure in Norway and so dirty in England?" Naturally, I could have referred to industrial pollution, the purity of the arctic fields of snow and ice, differences in wind regimes, etc., but we were all aware of Bergeron's meticulous studies of atmosphere turbidity. I thought the great Bjerknes would not be satisfied with anything but a really deep answer. Before I could find a suitable response, the professor answered his own question: "Because the Norwegians don't open their windows."

Vilhelm Bjerknes was always stimulating. Nothing was so trivial that he could not create interest around it. At the table, he told a few Swedish jokes which, apparently, he remembered from the years (1893–1907) when he held a chair in Stockholm. Since he was familiar with Freud's analysis of wit and classification of jokes, his stories took on a deeper meaning when seen against this background. He was particularly interested in colossal exaggeration as a distinct class—so typical of Swedish and American humor.[13]

After a few weeks Jack Bjerknes returned from his studies in

Switzerland, unpacked his papers, and continued his analyses. Jack was like that—quiet, unassuming, and full of purpose. Undoubtedly, his studies of data from observatories in the high Alps had given him new insight into the vertical structure of passing storms, but we heard very little about his findings. It became clear, however, that both Jack and his father were thinking of building mountain observatories in Norway. Jotunheimen, the central massif of the south, with nearly thirty peaks in the range from 6,500 to 8,000 feet above sea level, offered an abundance of choice.

At this time, young and venturesome Hans Ahlmann, then a junior professor of geography at Uppsala,[14] was planning research on the heat and water budgets of a typical glacier in Jotunheimen. Plans for the summer of 1924 included some meteorological work on the glaciers and a search for a peak as a site for a permanent observatory. During the winter of 1923–24 I had worked hard to study weather analysis in Bergen and to clear my study program in Oslo. Since 1915 I had spent every summer on the drill-grounds of the army. I was now more than ready for a change. Great was my delight when Jack suggested that I join Ahlmann as his meteorological assistant for the better part of the summer.

At this time another plum fell into my lap. One of the meteorologists in Oslo had been granted ten months leave of absence, and I was invited to fill his place as a forecaster, beginning when the Jotunheimen assignment ended. Though my early resolve to become an educated man had now receded into the background shadows, I felt, very clearly, that some kind of victory had been won. And to visit the peaks of Jotunheimen seemed attractive beyond all reason.

Ahlmann had rented a small hut close to Turtagrø Hotel, about 3,000 feet above sea level. We were responsible for our own cooking and housekeeping. Ahlmann had certainly seen many more cookbooks than I had and, as a result, I was far more willing to experiment than he was.

In those days no roads penetrated the Jotunheimen massif; cairns showed general directions, but footpaths were hard to identify. The tourists who came to Turtagrø were of a type quite different from that seen nowadays. Many were mountaineers of the highest order, waiting for weather suitable for daring climbs. Others were less experienced but no less keen. Age and skill divided the tourists into three classes: the elite, the up-and-coming, and the rest. Turtagrø was a mountaineer's paradise.

Some of the men from the farms in the nearby Fortun Valley were highly experienced guides. Their business depended very much on the weather. Ahlmann, with his full beard, looked highly impressive; all questions concerning the weather were referred to his junior prophet. All I could do was look at the sky, pause, and then express myself in cautious terms. Somehow, oracles seem to fit well into mountainous surroundings.

Gradually, we became well acquainted with the leading rock climbers. Some of them began to pay after-dinner visits to our hut, conveniently bringing their own refreshments. To me a rock climber was—and still is—a hero of a unique kind, a type overlooked by Carlyle.[15] A real climber is a lover of nature, silence, and solitude, and, above all, of danger; he trains his mind and body and seeks out tasks where safety is marginal and public opinion is immaterial; he neither trembles nor boasts, and he is generous with praise for his companions in the rope. More: He is choosy of his listeners.

No less interesting were the professional guides. In earlier years the most famous guides in Jotunheimen had been Ole Øiene and Ola Berge. Ole was now close to eighty and Ola a few years younger, but both were remarkably active and managed their affairs with great skill. In their early years both had hunted wolves and bears (mainly for protection) and wild reindeer (mainly for sport), but such attractions did not exist any longer.

Early in their lives their roads parted. Ola, who owned a large farm in the lowlands, built a hotel at Turtagrø and catered to summer tourists; his stories about mountaineering were centered on reckless climbing, accidents, and rescue work. On one occasion three highly competent mountaineers (A, B, and C) had set out to tackle an untried route. A, leading, slipped, fell past B, and landed on a narrow ledge below which there was a clean drop of several hundred feet. With B as a belay, C climbed onto the ledge and found A unconscious and badly hurt. He feared that A might regain consciousness and become unmanageable on this narrow ledge. B made his way to Turtagrø, where he arrived in darkness. Ola soon gathered a few men, arriving on the scene at dawn with a stretcher and other gear. As soon as they were within earshot, B shouted, "How are you?" and C answered: "Bored!" Because A was still unconscious, C had been sitting guard over his friend for seven hours while icy water dripped down his back. Indeed, a very monotonous vigil.

Climber A reached Turtagrø alive. Under a local doctor's first care, followed by a considerable amount of hospital engineering, bones and flesh healed. A few years later, A returned to Turtagrø to enjoy small climbs. We all understood that it might be tactless to reveal knowledge of the accident in his presence. And Ola was quite certain that the climb with the stretcher was his most difficult piece of mountaineering.

While Ola had taken his tourist business to the mountains, Ole and his dainty wife had made a comfortable living in the lowlands. Ole, a tall, handsome, and stately looking man, had built a small hotel near the river at Fortun. Here, in their younger years, the couple catered to a few tourists, some of whom were wealthy Englishmen. Activities included rock climbing, salmon fishing, reindeer hunting—anything in season and in good taste. In winter Ole would travel through Jotunheimen with horse and sled; in summer packhorses would do almost as well.

Many years before, one of Ole's most enthusiastic and frequent customers was a British army officer who later rose to exalted rank and was elevated to the peerage. He was interested in salmon fishing as well as reindeer hunting, and he took a great liking to companionable Ole. After a few visits the officer, who was then serving at Buckingham Palace, invited Ole to London and treated him royally. Ole returned to Fortun with a signed photograph of Queen Victoria. When Ahlmann and I questioned him about his impressions of the queen, he answered rather laconically: "She wasn't much of a woman; she was short and fat."

As snow began to fall in the high mountains Ahlmann and I packed our gear and returned to Bergen. All agreed that the summer's work had been successful and that the top of Fanaråken (7,000 feet) would be an excellent site for the new observatory.

After a few days in Bergen I went back to Oslo. I was now to be considered not only as an independent forecaster but also as one representing the Bergen creed. At that time Oslo was not in the forefront of meteorological innovations. Being a "freshman," I tried and succeeded in not getting involved in arguments that could not be resolved with reference to theory or observation. On the whole, the year was singularly uneventful. In my spare time I managed to accommodate a course in mechanics and write my master's thesis. The economic depression that set in after the First World War was now heavy and the authorities were anxious to reduce the number of civil servants; even salary reductions were actively considered. Fortunately for me, there was a vacancy in the Geophysical Institute in Tromsø and I was lucky enough to be selected.

About a month before my assignment in Oslo came to an end something unusual happened. A Soviet meteorologist, Professor E. Tikhomirov, arrived together with a certain Mr. Saltykov, affiliated with the Soviet Trade Mission to Norway. Like other foreign meteorological visitors, Tikhomirov was heading for Bergen, but had also obtained permission to stay in Oslo. For a period of three weeks the two of us worked side by side, part of the time in routine analysis and forecasting. I understood that Tikhomirov was not interested in forecasting as such; his concern was to explore how the new polar front and airmass ideas might be exploited in his research on atmospheric processes. Like Saltykov, Tikhomirov was a warm and unpretentious person, one of the rare individuals with whom I felt I would like to work indefinitely. It was clear that the Russian Revolution and hard times had left their marks. Both men walked about with their pockets bulging with Norwegian chocolate bars, which they consumed in unusual amounts. Even simple things had become promoted to luxuries.

Saltykov, who was especially interested in economic and social affairs, asked numerous questions, made copious notes, and seemed impressed with the high standard of living of the working class in Norway (which, in communist terminology, was a capitalistic country). On one occasion, when I mentioned that we paid so-and-so much income tax, he said that he would gladly pay twice that much at home for such blessings as we had in Norway. It was clear, however, that *they* were working (and suffering) for the benefit of coming generations. An era of ideologies and a new sort of *Weltgeist* were clearly in ascendancy.

As their visit was nearing its end, I was invited to some kind of reception at the Soviet Embassy. I cannot remember whether there were refreshments, but if there were the amounts must have been next to negligible. I had not had time to eat lunch and I grew hungry and tired as Tikhomirov spoke for three hours on end on some meteorological subject—in Russian. Madame Kollontai[16] presided. Judging from her facial expressions, she understood and enjoyed the discourse. Her face became especially lively when Tikhomirov mentioned Vilhelm Bjerknes and members of his entourage.

After the presentation Madame Kollontai expressed her gratitude for the help I had given the visitors. Then, she and a counselor engaged me in small talk, ending with an invitation to dine with them, Tikhomirov, and Saltykov at the Grand Hotel. Madame Kollontai, then about fifty, was a lady of immense charm, and the party was most enjoyable. The conversation was informal, light, and witty, with an undertone of idealism. Until then, I had thought of communism as a system of dogma within which thinking was reserved for the top of the pyramid, with the remainder serving as a resonator. To me communism was an intellectual straitjacket, a mere formula, dressed in fine words. Nevertheless, I found myself enjoying the company of a group of people as soulful as I could ever wish to meet. I left the party with a feeling of having observed a conflict between *formula* and *soul,* and it seemed to me that the Russian soul would outlive the brutality of the communist regime.

As my Russian friends left, the time had come for me to pack and go north. Fourteen years had passed since I left *Nordland*. I was anxious to reach Tromsø while the midnight sun was still there. Another six weeks and the birds would be moving south.

Chapter 3

In the Grip of the Arctic

I will go back to the Northland,
For the Northland is calling to me.
—Robert Service

I arrived in Tromsø while the sun was shining day and night. I soon found that Tromsø, with only 10,000 inhabitants, was a cultural center out of all proportion to its size. As the "Capital of the North" and the "Gateway to the Arctic" the town had attracted a number of institutions and a remarkable concentration of professional men. Naturally, intellectual activities related to the Arctic were much in the foreground; geophysical subjects, medicine, archeology, and anthropology were appreciated in wider circles; local patriotism was fashionable; and influential men were already discussing the possibility of establishing a university.

The Geophysical Institute, under the direction of Professor O. A. Krogness, was responsible for the weather services not only for northern Norway and adjoining fishing fields but also, and in a wider sense, for the whole sector of the Arctic within which Norway had an interest. The Institute operated meteorological radio stations on Bear Island, Spitzbergen, Jan Mayen, and the east coast of Greenland. It also provided information to seal hunters and other interested parties.

One of our neighbors at Tromsø was a music teacher. Fairly soon after our arrival she happened to mention that the recent fisheries had been so bad that she had hardly any students. Realizing that I did not see the connection, she added that when the fisheries failed, some people could afford only necessities—and others less. Many could not pay their bills; the merchants, the dentists, the doctors and even the lawyers had little to do; and music was lowest on everybody's list of priorities. When the fisheries failed, the heart of the economy did not pump enough blood through the system. On the other hand, when the fisheries were good, the pump was overly busy. The typical *Nordlander* lives in gusts and lulls.

The pulse of the economy was irregular; it depended not only on the major seasonal fisheries but also on the catch of seal in the "east ice" and the "west ice"—the White Sea region and the Greenland ice belt.

The sealing business has its ups and downs. The seal pups are born on the ice; their heads are heavy and they need to put flesh and fat on their bones and learn a few tricks from their parents before they are able to swim. Any storm that plays havoc with the ice may kill off pups in enormous numbers.

The pulse of the economy depended to some extent too on the halibut fisheries on the Bear Island banks, but to a much larger degree on coal mining on the west coast of Spitzbergen and on hunting and other activities in the whole Spitzbergen region. Some of the subprofessional staff at the Geophysical Institute had served on the arctic stations, where they could combine hunting with their regular work; fox trapping was particularly rewarding.

I soon found that my colleagues and the other Spitzbergen hunters were remarkable men, and I began to pick up some phrases from their rather unconventional language. A "blood place," I discovered, is a large lump of frozen blood and snow marking a place where a polar bear has rested while trying to escape after having been shot by a spring-gun.[17] While he rests, blood freezes over the wound and he gains strength to continue. However, body heat soon melts the ice, and a red trail leads to the next blood place. The track tells its story. Two or three places means a fine hit and little work for the hunter; twenty places means either a poor hit, or—perhaps—a heroic bear deserving of admiration. Anyway, the hide is added to the catch. "Warm food" may either mean meat of some kind boiled in a pan in the hunter's shack, or warm blood drunk directly from any animal just killed. The dread enemy of the wintering hunter is scurvy, and all know that anything fresh—blood or meat—is wholesome. "Spitzbergen tea" is made of the contents of the crop of a ptarmigan mixed with snow and brought to a boil. The ptarmigan digs through the snow, particularly in windswept places, and consumes leaves of plants rich in vitamins. The hunters know that scurvy is a slow killer, and, if it comes late in the winter season, it may lose its race with the sun.

On one occasion a sturdy and reckless hunter returned from a winter in the Arctic, concealed the evidence of a beginning scurvy, and enlisted for a year's service at our station on the east coast of Greenland. Well before Christmas a state of scurvy was clear for all to see, but the Greenland ice was already solid and evacuating the patient was out of the question. Though the Tromsø doctors were in no doubt about the outcome, our director, Krogness, took time off to read all the literature on vitamins that he could lay his hands on. After a week or two, Krogness sent a radio message to our foreman in Greenland to put large quantities of dried peas to soak, and then to shave off the sprouts and feed them to the sick man. This diet was to be supplemented with "fresh food" and any variant of Spitzbergen tea that might be concocted. The scurvy yielded, and next summer the man returned to good shape. In the early twenties, not much was known about vitamins and the Tromsø doctors were full of praise for Krogness—an expert on earth magnetism and northern light. An unusual imagination is more important than knowledge.

In my childhood I had taken life in *Nordland* for granted. Though *Nordland* is but one of the counties of northern Norway, the name is commonly taken to refer to the region as a whole. Returning to it, I tried to make a deliberate effort to observe and understand what I saw. A man of great influence in Tromsø at this time was Just Quigstad, a scholar in many fields and an outstanding authority on Lapology. Quigstad had started out as a student of divinity, but the history, culture, and folklore of the Lapps had gradually become an absorbing interest. Though my senior by almost fifty years, he took an interest in my work and was always helpful with information on bygone times and the cultural background of the North.

Nordland is a narrow sector of the arctic region. Over long spans of time cultural influences travel great distances; the Lapps might have received many impulses from sources other than Norway and Sweden. About one thousand years ago Finns moved in numbers from the Baltic region into the present Finland, and many hunters intermingled with the Lapps who, by then, had taken to reindeer herding.

Quigstad's interests spanned a wide geographic area. Across the Norwegian Sea, in Greenland, the Eskimos lived by hunting and fishing, much as the Lapps must have done before they turned to reindeer capitalism. Was there any connection between the two cultures? When a young man, Quigstad decided to find out. In preparation for a visit to Greenland he studied the Eskimo language from such books and documents as he could get hold of. Pronunciation was a great problem; would he be able to make himself understood to a common Eskimo?

Eventually Quigstad took ship from Copenhagen, arrived at Julianehaab and, to his delight, saw a young Eskimo passenger coming aboard. Here was a golden opportunity for Quigstad (Q) to practice on an Eskimo (E).

"Good morning!" said Q.
"Good morning!" answered E.

Quigstad was overjoyed; he had made himself understood. Constructing a new sentence, he continued:

"Where are you going?"
"To Ivigtut," answered E.

It all sounded very simple, and Q thought out a third question:

"What are you going to do there?"

And the young Eskimo answered:

"I am going to kill my father."

A mistake somewhere, surely! Q looked in his books and notes, rephrased his question, and received the same answer—again, in a note of reverence rather than malice.

The explanation was reasonably simple. At that time, when an Eskimo reached such an age that he felt that his life had become a burden to himself and his community, he decided that the time had come to depart. It was the duty of the eldest son to officiate, observing the time-honored rules and customs handed down through generations. In all aspects the act was charitable and solemn—and who could be better qualified and act with greater dignity than the firstborn son?

As far as I can now remember, Quigstad found no tangible evidence of any connection between the Lapp and the Eskimo cultures. In particular, the Lapps seemed always to have cared well for their elders and managed to provide for orphaned children. On the other hand, there was much well-known evidence of cultural influences between the Lapps and their neighbors to the east—the Kola Lapps, the Samoyeds, and, perhaps, even more distant tribes. Neither the Lapps nor the Eskimos had been seafaring people, and the Norwegian Sea had served as an effective barrier.

However, Eskimo influences had spread westward, across the Bering Strait and far into Siberia—how far no one could tell. It is known that the Chukchi people and other tribes along the coast of Siberia help their elders out of the miseries of old age. To them the northern lights are the souls of their forefathers enjoying reindeer races and similar sports. But no one can say that such traditions have been imported from the Greenland Eskimos. Life in the distant Arctic is hard. Local customs often reflect natural responses to real needs. Everywhere and at all times, what is said to be cruel or inhuman is just that which is not strictly necessary. Among the civilized intruders in the Arctic it has long been an unwritten law that consideration of an individual shall not stand in the way of the safety of others. And cannibalism is certainly more prevalent among civilized men than among the arctic natives.

Though in modern times tools and practices have spread through much of the arctic belt, folklore elements have proved to be far more conservative. Often it is difficult to decide whether cultural similarities have come from actual contacts or from natural responses to an environment that is fairly uniform and exceptionally demanding. The "voices" of the mountains, of the sea and ice, and of the vaults above are likely to have left similar marks in the collective memory of the different races. And for the people of the subarctic zone, nothing could have been more mysterious and impressive than the infinitely varied displays of northern light.

My beliefs were confirmed in 1948 when a Hindu scholar in the Asiatic Institute in Calcutta showed me leaves of the palmyra on which had been inscribed references to the northern lights. Aurorae are not seen in India, though on some rare occasion a faint patch of light may have qualified for inclusion in the class. But such faint and exceedingly rare occurrences are unlikely to leave folklore traces. The kind scholar, citing archeological evidence in support of this view, explained that at least one of the many invasions of India through the Khyber Pass had come from the White Sea region. The palmyra record probably testifies to the immense importance that the early arctic people attached to the mysteries of their sky.

Though the cultural aspects of arctic life and lore were fascinating, there were also practical problems that required my attention. The Institute was starved for funds and grossly understaffed for the many duties that had to be met. Fishing in northern Norway goes on throughout the year. The autumn storms begin early. The heavy concentration of storms, fisheries, and trade during the Lofoten season (early January to early March) was only five months off. My two colleagues had the advantage of several years of forecasting experience in the north; I could supplement this with knowledge of the more scientific techniques developed by the Bergen group. The fishermen started out early in the morning, fished all day, and needed the weather forecast before the telegraph stations closed at 7:00 P.M. Unless our storm warnings were received by that hour, the fleet might be caught on the banks the following day. On the whole, things went quite well, and the season came to an end with general satisfaction. Once in a while I wished my grandfather had been there to see how things had changed: better boats and a reasonably reliable storm warning service.

The Lofoten season over, a new problem came my way. Kautokeino, one of our precious stations far into Lapland, needed a new barometer. A mercury barometer is a rather delicate instrument and must be moved with great care. Two years earlier a colleague had succeeded in transporting a barometer to Kautokeino, driving with reindeer and pulka sled [a traveling sledge shaped like a boat] through the wilderness—a journey that took four days in good weather. Unfortunately, he had slipped on the icy doorstep of our observer's house and had broken the instrument. Now I was asked to replace it. I needed no persuasion.

I took the coastal steamer to Bosekop where the annual Lapp market was in progress. There I was lucky to find the brother of our observer in Kautokeino and made arrangements with him. He was going to drive home a *raide*, a string of alternating reindeer and pulkas, loaded with wares of different kinds. His seventeen-year-old daughter would serve as my guide, and he would meet us every evening at the posting stations along the route.

I soon discovered that the Lapps have but little sense of time. To be in a hurry or to show indications of impatience is taken as outright rudeness. When I asked my factotum[18] how early we could start next morning the answer was 6:00—or 7:00, or 8:00. At about 2:00 P.M. he began to gather in the reindeer; we took off shortly before 4:00 P.M. and the sun set at 6:00 P.M. Being interested in sightseeing, I brought up the question of timing, and probably overemphasized the virtues of an early start. My good man listened politely and patiently. When I was through he responded with the wisdom of a Lapp philosopher: "If you are that busy you should have started yesterday."

The guide was a very intelligent girl, skilled at reindeer driving, polite and well mannered. She seemed to be rather hazy about world geography, but she knew every detail of the terrain within a strip of a few miles along our track. One moonless night she left me and made a diversion into some clumps of dwarf birches while my reindeer carried on. After a while, when I was beginning to worry about my own safety, she reappeared with a small parcel that she had hidden in the bush two weeks earlier on her journey northward to the marketplace. As I drove on I began to appreciate the habits and problems of the Lapps, their sense of wit, and their friendly way of life. It is the reindeer, rather than the Lapps, that set the tempo. The animals feed on reindeer moss; since it grows very sparingly the deer have to search over large areas. They get at it by scraping through hard and crusty snow. Feeding is a time-consuming occupation, and after four days of relatively hard driving, our animals were tired out.

While I attended to my business in Kautokeino, my factotum drove to Oskal, near the Finnish border, and returned four days later with fresh animals. On the homeward journey we ran into a howling storm, with falling or drifting snow blowing into our faces. I was not certain who found the way, the guide or the reindeer. As we approached the coast the snow became very deep and the going hard. Before noon of the last day we had to abandon one of the reindeer. Late in the evening we arrived at the house of the commandant of the Finnmarken region, an army officer whom I had known well since my school days in Trondheim. The guide was offered shelter for the night but declined, having to care for his tired animals. Next the commandant poured sherry, and the Lapp emptied three large glasses in rapid succession. When the empty glass was held out again, the commandant ignored the hint. I thanked my guide warmly for his good care and sent greetings to his daughter. Then we sat down to a dinner that is still lively in my memory.

Upon my return from Lapland I had before me a long stretch of forecasting duty, which would have been routine had not an expedition "over the top of the world" intervened. Involved in the enterprise were two strange personalities—Roald Amundsen and Umberto Nobile—whose early and fruitful collaboration developed into jealous competition and ended in disaster.

Amundsen, the grand old man of Arctic and Antarctic exploration, was physically robust and in most respects a man of the Norse sea-king type—a leader and a man of action. He had been the first to visit the magnetic North Pole, sail the Northwest Passage, and plant a flag on the South Pole. He had sailed, as had others before him, the Northeast Passage, and in 1925, using aircraft, he almost reached the North Pole. In 1926, his main interest was to pave the way for air communication across the Pole, between Norway and the Bering Strait and beyond. Nobile, a physical weakling who had failed three times to qualify for ser-

Fig. 3.1: Finn Malmgreen (d. 1928), as drawn by Knut Knaus in 1925.

vice in the Italian army, has justly gained recognition as an able and imaginative aeronautical engineer, a designer and builder of semirigid airships—just what Amundsen needed to realize his dream.

Amundsen found his way to Nobile, and in early 1926 a dirigible, the *Norge*, hoisted the Norwegian flag. Though Amundsen led the expedition and his wealthy American friend, Lincoln Ellsworth, footed many of the bills, Nobile was to captain the ship, with Hjalmar Riiser-Larsen, a Norwegian naval officer, as navigator and second-in-command. A fifth person of interest was the Swedish meteorologist Finn Malmgreen (Fig. 3.1), who had served under Amundsen on the *Maud* expedition (1922–1925) to the Arctic. His assignment in the *Norge* was to be the expedition's meteorologist. As it happened, the three men, Amundsen, Nobile, and Malmgreen, became the main actors in a tragedy that was played and replayed in 1928.

The airship *Norge* took off from Rome on April 10, 1926, stopped at Pulham (Norfolk), Oslo, Gatchina (Leningrad), and Vadsø, arriving at Kings Bay, Spitzbergen on May 7. Also at Kings Bay at that time were Commander (later Admiral) Richard Byrd and his famous pilot Floyd

Bennett, getting ready for their attempt to reach the North Pole. Since neither Amundsen nor Byrd was averse to publicity, Kings Bay soon attracted the attention of the mass media around the world.

On the flights from Rome to Leningrad, Malmgreen had obtained forecasts from different weather services, used his own judgment, and provided good advice. From Leningrad to Vadsø and onward to Kings Bay and the Bering Strait, the Tromsø Geophysical Institute provided forecasts for Malmgreen to use in his briefings of Amundsen and his lieutenants. In response to our advice the stop at Vadsø was shortened; the journey was uneventful and the ship arrived safely at Kings Bay. The major problem was the long journey from Kings Bay across the North Pole to Nome, Alaska, where elaborate arrangements had been made for Amundsen's arrival. For a few days the large-scale weather situation remained ideal, but repair work on the ship delayed Amundsen's departure. Byrd and Bennett, however, were able to take off and, after a reasonable number of hours, they returned, claiming to have reached the North Pole.

In the meantime, a disturbance in the form of a low pressure area had moved eastward into Siberia. Though reports were very sparse I had reason to believe that the disturbance was heading for Alaska. In spite of the many uncertainties, the weather in the Bering Strait and adjoining areas seemed to be deteriorating.

The ship was ready to take off on the morning of May 11. With the planned maneuvers around the Pole, the dropping of the flags of Norway, the United States, and Italy, and other appropriate formalities, the flight might take as much as seventy hours. The final forecast for the flight stressed three significant aspects. One, a large high pressure area stretching all the way from Novaya Zemlya to the Canadian prairie would ensure fine conditions above the eightieth parallel. Two, a low pressure system moving in from East Asia was likely to cause clouds and strong winds in the Alaskan approaches. And three, favorable weather would prevail from the Pole to the Canadian plains as far south as the fiftieth parallel.

In the twenties publicity, with income from press and radio, was an important consideration, and Amundsen was not inclined to change his plans. Nome, with arrangements for a reception, remained his destination. They took off at 10:00 A.M. To begin with things went well, the Pole was there "for all to see," flags were dropped, and Ellsworth's birthday was celebrated. It is hard to imagine the stirrings in Amundsen's mind as they circled over the Pole. His life had been spent in polar exploration, and there was but little left now to be explored, though perhaps enough for a younger and weaker generation. Amundsen was a proud man.

They flew on toward Nome, but as they approached the Bering Strait, the winds freshened, and clouds made navigation difficult; there were high mountains to be avoided. The storm center was now south of Nome. The winds became strong from the northeast, and ice began to accumulate on the canvas. The ship did not quite reach Nome. An emergency landing was made; very skillfully, Nobile set the ship on the ice with a minimum of damage, and no bones broken. Amundsen had reached Teller, not far from Nome. A little less than seventy hours had elapsed since the *Norge* took off from Kings Bay; more than three days had passed since the forecast of the storm was prepared. Until then, I had never attempted to forecast for such a long period.

Immediately after the *Norge* flight, Amundsen, with contributions from Ellsworth and Riiser-Larsen, published an account of their expedition. In it we find the soothing statement, "We never heard an angry word or saw an unpleasant expression during the entire flight." From other accounts one gains the impression that the flight was filled with sharp exchanges. Amundsen, then fifty-four, had accumulated a long record of unsurpassed feats in polar exploration by conventional means, but had little useful knowledge of dirigibles. Nobile, on the other hand, knew all that there was to be known about dirigibles but had hardly any understanding of the psychology and mechanics of polar exploration. There were also other differences. Amundsen was a leader of men; he knew how to command and he knew when he had been obeyed. In contrast, Nobile knew neither how to command nor how to obey; he was a professor in a general's uniform who thought that the solution to all problems outside his own narrow field would arise through discussion.

Amundsen and Nobile had at least one thing in common: personal vanity, which sometimes influenced their actions. On the *Norge* flight all luxuries were supposed to have been left behind to save weight and gain range—in case of need. Nevertheless, Nobile and his officers had seen fit to carry dress uniforms in their personal luggage for use at the landing in Alaska, while Amundsen, Ellsworth, and Riiser-Larsen had to appear dressed more or less like Eskimos. The relationship between Amundsen and Nobile became further strained when Nobile began a tour of the United States with popular lectures on his transpolar flight. By any standard Amundsen qualified as an elder statesman, though on occasion, and in relation to Nobile, he may have appeared to act more or less like an aging prima donna.[19]

Even before we began to work on the Amundsen expedition, Krogness had been asked by an American colleague if we could accommodate measurements of solar radiation in our Spitzbergen program; if so, instruments would be provided without cost. The Institute being starved for funds, the response was a foregone conclusion. It became my assignment to install the instruments at Kings Bay. We had no skilled observer there, but other things seemed favorable. The doctor in this little mining town (who was not overburdened with duties) had indicated his willingness to attend to the observations.

I arrived in the middle of June, on a steamer that came to fetch coal from Kings Bay, the first one since the previous autumn. I had completely underestimated the importance of this first arrival, for the ship brought fresh supplies: meats, vegetables, fruits, fashion magazines, a variety of bottles, and, inevitably, the annual wave of the common cold. The doctor saw nothing but trouble ahead of him.

The director of the mines and his wife looked at things in a different manner: The occasion called for a celebration. After dinner, and with everyone in high spirits, one of the engineers suggested that we send a telegram with good wishes to Roald Amundsen. Not knowing his address on that particular night, messages were sent to many plausible places. As I heard later, almost all of them reached him as he was touring the United States. About 2:00 A.M. our host invited the men to his sauna. This was my first experience with excessive heat. Thoughtlessly, I said I'd rather go for a swim in the bay. When I was pressed to do what I had said, we went down to the shore, and seeing a small berg a few yards away, I could not resist the temptation to touch it. The night was sunny and perfectly calm. The chill was not frightening—unless one stayed in too long.

The doctor was an elderly German or Austrian (which, he never made clear) who had obviously seen better days. Toward the end of the First World War he had served as a medical officer at Przemyal, a delousing camp in Poland. The number of Russian prisoners was very large, and the average number of lice per prisoner was simply unbelievable. The doctor had nothing against the Russians; on the contrary, they were simple and peaceful people who had to suffer for the mismanagement of the czarist regime. Toward the end of the war the food situation deteriorated very badly. The Russian prisoners—with or without lice—were a low priority lot. One day the prisoners went on strike, refusing to obey orders. The officer in charge was summoned. With the aid of an interpreter, he asked, "What do you want?" He was rather amused when the prisoners' spokesman said, "If you not give us bread, we not give you lice." Apparently the basis for negotiations was too flimsy, and the strike collapsed.

Although earlier, in response to our radio message, the elderly doctor had declared himself willing to attend to our instruments, it transpired that in doing so he had assumed that his duties would be minimal and that his nurse could do the actual work. When I pointed out that things were not quite that simple, he insisted that he could not be tied to any routine. Failure of my mission was clearly within sight.

In a last effort to win him over, I opened a bottle of Cluny which I had with me in case of need. After a glass or two he forgot his troubles and became quite lighthearted. He told me that if he were young again, he would settle down in Spitzbergen and develop the land into one of the major tourist centers of the world, something far more attractive than Switzerland. Hunting as a sport—not as a miserable living—would attract tourists in all seasons, as would winter sport almost without interruption and rock-climbing in the midnight sun. With proper accommodations, sun-worshippers from Siam and other countries would come in flocks. To balance the trade he would invite Muslims to celebrate their Ramadan during the winter darkness. Among the many restrictions imposed during Ramadan he emphasized especially abstention from sexual intercourse during daylight hours. He laughed heartily at his bright ideas, and ended with his usual: "*Bin jung gewesen; bin alt geworden*!" ["Once I was young; now I am old!"]. Still, he was quite certain he would have nothing to do with my instruments.

In the meantime, the ship on which I had arrived had left with a load of coal. My next opportunity to depart was eighteen days off. Birdwatching was an obvious pastime. In my childhood I had become familiar with a great variety of migrating birds. Here, at Spitzbergen, I was lucky enough to gain some insight into their social habits. Near the town I had often seen a female eider duck that had a deformed wing. Some of the primary feathers had become deranged and I doubted that she could fly. Here was a unique bird that I could identify from day to day. For reasons I do not wish to explain, I called her Anna.

One day as I was walking toward a bird cliff, I inadvertently frightened a mother eider duck off her nest. Only two of her three chicks were able to run with her. Left behind was a helpless creature—just skin and bones. After a few minutes fulmars and other birds discovered the unprotected nest. With a peculiar sense of guilt, I decided to stay to protect the chick, hoping that the mother would soon return. Birds, however, are notoriously poor at arithmetic. After a while there is but little difference between two chicks and three. Then, to my surprise, Anna came along, took possession of the nest, and behaved as if she had just found her dear little baby. Two days later I saw Anna swimming about proudly with her stepchild. Her spinsterhood was no longer evident—except for those who remembered her damaged wing.

Though bird-watching became my main pastime, I also had occasion to visit glacial tongues projecting far into the sea, observing how large bergs break off from the mother glaciers. Plant life was another attraction. Though the number of species was quite small, they all seemed highly sensitive in their responses to variations in the environment. I also had occasion to visit the coal mines and to travel on the world's northernmost (and probably shortest) railway. Eventually, the ship arrived and I returned to Tromsø—with my instruments.

At the very end of August 1926 snow began to fall. Next morning the depth was close to thirty inches. Tromsø was a skier's paradise; I simply lived on skis. In my daily work I found the rucksack more convenient than the briefcase. Late in March I joined a party of four on a skiing expedition, zigzagging through the rugged mountains between Tromsø and the Finnish border, continuing on the flats southward to Keinovoipio in Finland. Next, we crossed the wilderness of Sweden that separates Finland from Norway, and then skied southward, reentering Sweden, finally arriving at Abisco, a winter resort on the Stockholm-Narvik railway. The route we had chosen was rather unusual and generally considered unsafe, mainly because snowstorms are frequent in the mountains and shelter hard to find.

At Keinovoipio we needed to replenish our supply of food, but language difficulties stood in our way. Among the four of us we could make ourselves understood in a number of languages, but not in Finnish: Fifteen cases, postpositions, forty-letter words and a *Kalevala* intonation were more than we could handle. [20] One member of our group, a clergyman conversant with Latin and Greek, had learned sign language in his youth, but even this proved useless.

Then came an inspiration. Keinovoipio had a public telephone. Next to it I saw hanging four ptarmigan in their snow-white winter plumage. Two hundred and fifty miles farther to the south was Haparanda, a town situated right on the border between Finland and Sweden. Surely, the telephone operator there would speak both languages. So I tried: "Hello . . . I am here at Keinovoipio and next to me stands Mrs. So-and-so; she has four ptarmigan which I wish to buy and have cooked to take with us on our long skiing trip tomorrow. We are very short of food and need the ptarmigan very badly. . . ." Though the general idea seemed simple enough the transaction became rather complex. It took Miss "Haparanda" some time to understand the geometry of the situation and for Mrs. "Keinovoipio" to decide whether she wished to part with the birds. Thereafter the two ladies engaged in a lengthy conversation, discussing various aspects of ptarmigan cooking. The telephone system was a one-line affair and before long women from different places joined in the discussion. One of them asked to speak to me, just to hear my voice. There seemed to be no end to it, and I began to fear an enormous telephone bill, but their system of accounting was sufficiently flexible to make allowance for circumstances. The bill was very reasonable, and so was the cost of the birds. And best of all, the wives of a dozen farmers and hunters had experienced something unusual and had acquired a topic of conversation to which they could return over and over again.

The ptarmigan came in handy the next day. A storm with dense drifting snow slowed our crossing of the high mountains. After nineteen hours of difficult skiing we reached the nearest farmhouses in Norway—at four o'clock Good Friday morning.

The heavy snowfall during the winter delayed spring considerably. On the following May 20, the first day when the sun does not set, the snow was patchy. Nevertheless, I found it convenient to use my skis to reach a nearby hill from which I could watch the sun's rim as midnight approached. Though the winter had been unusually long it had certainly not been unpleasant. On June 19, as I was driving from Vadsø to Nyborg in Varanger to prepare for measurements of earth magnetism during the forthcoming solar eclipse, remnants of old snowdrifts were still bothersome. Within less than a week summer had replaced spring. The snowdrifts had disappeared and the meadows turned golden with subarctic varieties of dandelions, buttercups, and similar flowers. A broad current of hazy air, with its origins in the plateau north of Pakistan, had invaded Lapland, bringing with it midday temperatures above 90°F. In the Northlands, summer does not develop; it just arrives.

The solar-eclipse work was routine. My job was to have the instruments mounted and functioning by the day before the eclipse, when Krogness would arrive and take charge. Within 300 yards of Nyberg was a German expedition headed by Professor Franz Linke and staffed by two of his doctoral assistants. Their instruments and equipment were of the *dernier cri* ["last word"] type while ours could best be described as being of the *dernier ressort* ["last resort"] variety—sturdy, homemade handiwork, but not to be found in museum collections.

Quite unexpectedly, J. W. Sandström turned up. Though traveling as a private person, he was often referred to as the Royal Swedish Solar Eclipse Expedition—since most things of consequence in Sweden are designated as royal. Sandström was a person of consequence. As a young lad he had been educated in a small smithy in the extreme north of Sweden. Before he was twenty he had been identified as a local genius by a road-building engineer who offered to train him to become a foreman. Sandström declined the offer, saying that if he could not become a physicist he would rather remain a blacksmith. As he matured, his gifts became widely recognized within his parish. Some of the better-off neighbors collected funds and managed to get him accepted as a special student of science in Stockholm where Vilhelm Bjerknes then held a chair in fluid dynamics. On arriving in Stockholm, Sandström soon forgot about science and became absorbed in Russian. However, the terms of his admission had to be met, and Bjerknes agreed to supervise his study program. Not long thereafter Sandström became one of Bjerknes's junior collaborators.

While working in Bergen I had heard a great deal about Sandström, his ingenuity as well as his antics. Sandström was a genius and took his cue from simple things that he saw in nature. Once when skiing in the Swedish mountains he saw what many others must have seen—that when air is cooled it begins to slide downhill. But Sandström went further, considered the deeper reasons, and came out with a fundamental theory, a law of nature, which states that "to maintain atmospheric motion against friction it is necessary that the air be heated at high pressure and cooled at low pressure." Many great scientists tried to disprove Sandström, and it was only after Bjerknes had given academic varnish to the blacksmith's creation that the theorem became generally accepted.

Sandström was not only a genius but also a child of nature. Although he progressed in science, his social manners remained more or less as they had been when he worked in the smithy. When he had risen to a position of consequence in Stockholm, problems of protocol arose, some of them seemingly insoluble. On one occasion, Mr. and Mrs. X felt that some recognition of Sandström was unavoidable and that a minimum of calculated risk would be achieved by inviting him to lunch, without other guests, at their home. After the meal, the host was called to the telephone, the maid went to fetch coffee, and Mrs. X was left alone with her guest. Very soon Sandström, good-natured as usual, placed his hostess on his knee and engaged her in small talk. Before she could decide how to escape gracefully, the maid returned with the coffee tray. Somehow Sandström managed to get the maid seated on his other knee, and a good-natured conversation of expanded scope followed. Soon Mr. X returned. Before he could decide what remark, or action, would be appropriate, Sandström, with his disarming smile, said, "You are a lucky man to have *two* such women in the house."

This happened many years before the solar eclipse of 1927. At Nyborg, Sandström's amorous advances were far more sophisticated. It so happened that our hostess was much younger than her husband, and Sandström was an astute observer with an inventive mind. On one occasion he borrowed my nail scissors, cut a button off his waistcoat, and called on our graceful hostess to come to his room and sew the button on without him having to take the vest off. When I returned from work late that evening I was told by a reliable source that the button had been replaced by one of the maids sitting, with the others, in the farmhouse kitchen.

Our unorthodox instruments required frequent attention and Krogness and I worked through the night to make sure that everything was in good shape for the early morning eclipse. Neighbors from a wide area had gathered at Nyborg not only to see the eclipse but also to watch scientists at work. Things like these do not happen often in Lapland. Sandström had no instruments to attend to. Shortly after midnight he had his huge thermos flask filled with steaming hot cocoa, borrowed a warm blanket, and walked to the top of a treeless hill. There he made himself comfortable and enjoyed watching the spectacle. A few weeks later Krogness and I were thrilled to read a popular article on what he had observed.

After a lull in 1927, activities in arctic exploration flared up in 1928. The first event was the flight by Sir Hubert Wilkins and Carl Ben Eielson from Alaska to Spitzbergen. We in Tromsø did not know of the flight until a few days after the landing somewhere in Spitzbergen; the "somewhere" turned out to be near Green Harbor. On April 15 they had taken off from Point Barrow, heading for Kings Bay, a flight that was planned to take twenty hours and span 171 degrees of longitude, or about 2,500 miles. About twelve hours before takeoff a storm of the Bjerknes type began to develop right on the east coast of Greenland, about 71°N. At the time of takeoff intensification set in. In all important aspects, the storm resembled the so-called Ryder cyclone discussed in the previous chapter. The storm reached peak intensity on the west coast of Spitzbergen just as Wilkins and Eielson arrived. A disaster was averted when Eielson, almost blinded, set the plane down in deep snow. Five days later the storm cleared and the aviators found their way to the radio station at Green Harbor, from which the good news of their safe arrival was spread far and wide. Wilkins and Eielson arrived in Tromsø on May 15. Since May 17 is Norway's independence day, their presence added much color to the celebrations.

Wilkins spent much time at the Geophysical Institute. Since explorers are in the habit of dramatizing the manner in which they have had to brave the elements, he felt that he had to document his case, which, indeed had been an extreme one. He asked Krogness if we could prepare a factual report on the storm. It became my assignment to write it up. What he needed was dignified publicity to assist in raising funds for new enterprises. It was understood that my factual report would be used by Sir Napier Shaw, a prominent and exceptionally eloquent British meteorologist and physicist, to write an article of the type Wilkins had in mind. My weather charts told their own story and there was but little need for a text. Nevertheless, I wrote as well as I could, and tried to give Sir Napier a few points that might prove helpful to a person who had not specialized in forecasting; so often the beginning is the most difficult part. My effort proved completely unnecessary, for Sir Napier started off by paraphrasing, not Sverre Petterssen, but Julius Caesar: "*Atmospheria est omnis divisa in partes tres*. . . ." For me it was difficult to connect the *partes tres* with anything related to the storm development. Perhaps Sir Napier, too, had had his misgivings, for in a second article on the same subject we find no reference to Caesar.[21]

While Wilkins and Eielson were touring Norway, receiving the acclaim they had so richly deserved, an Italian expedition, sailing in the new and supposedly much improved dirigible *Italia*, was on its way to Spitzbergen. The expedition was commanded by Nobile, now a general. Its program included a visit to the Pole, exploration over a wide area, and a series of geophysical measurements. Wisely, Nobile had secured the services of Malmgreen as his meteorological advisor. The arrangements for the provision of forecasts were about the same as for the *Norge* expedition two years earlier.

The political wave that carried the *Italia* expedition was Mussolini's desire to show the world that Italy was leading in the air and could do as well as others in arctic exploration. Two years had passed since the *Norge* flight; aviation had advanced, and many were anxious to see what Nobile could achieve in his new ship. Though all wished him well, some were skeptical about his ability to lead and manage. We in the Geophysical Institute were among the skeptics, and it was fortunate for us that Krogness was so generously endowed with wisdom.

Weaknesses became evident even before *Italia* reached her base in Kings Bay. Though Nobile had the general support of Mussolini, he appeared to have overlooked the importance of cultivating liaisons with the influential men on whom he depended for practical arrangements. Foremost among these was General Italo Balbo, who had little confidence in lighter-than-air craft and feared that his position as Italy's leading airman might be threatened by the upstart, Nobile. Dictatorships are remarkably productive in intrigues. The reins of bureaucrats were strong and well centered in Rome, and Nobile was naïve and trusting. As a base for his expedition he needed a ship to be stationed at Kings Bay where a hangar, mooring mast, and other facilities had been maintained since the *Norge* flight. The authorities in Rome had chosen the hopelessly outmoded and ill-equipped steamer *Città di Milano* as mother ship and arranged to have her commanded by Captain Giuseppe Romagna, a bureaucrat who was willing to take his orders not from Nobile but directly from Rome. The complications resulting from these and other arrangements were not just symbolic or formal; they were real and glaringly apparent.

On her way to Spitzbergen, *Italia* stopped at Stolp (in Prussia) for rest and repairs, and Tromsø began to provide Malmgreen with forecasts. On April 30 we sent him an appraisal of the general situation. Although conditions were not ideal, departure was recommended. The takeoff was delayed for two days, however, by local conditions at Stolp. The ship had to stop at Vadsø and, as a result of the delay, the winds there became a major concern. In the early hours of May 3, *Italia* left Stolp, the decision being based on our advice.

Though direct contacts with the ship were sparse, we managed to track her position by listening to radio broadcasts, realizing that some of them might be inaccurate. *Italia* lost some valuable time in ceremonial maneuvers over Stockholm, after which she continued on with much-reduced speed. At the northern end of the Bay of Bothnia, she seemed to sail eastward, or even southeastward, as if she were just following the coastline. However, she soon recovered a northward course. About midnight we reexamined all evidence and conferred with Krogness. As a result, a message was sent to Malmgreen recommending increased speed to reach Vadsø before strong winds would hinder mooring. Our advice was followed and *Italia* reached Vadsø on the morning of May 4. Soon after *Italia* landed, the wind increased beyond acceptable limits, and the ship had to hang on to the mast until next morning.

The weather was fine on the first half of the flight to Kings Bay, and marginal thereafter. However, on May 6, shortly before noon, *Italia* broke through the cloud deck; for the first time she saw *Città di Milano* and exchanged messages with Captain Romagna. When *Italia* arrived over Kings Bay, Nobile requested a ground crew to handle the landing ropes. Romagna refused assistance, saying that his men were not duty-bound to work on land. Anyway, he could not act without orders from Rome. This awkward situation was alleviated by men from the nearby Norwegian coal mines who handled the ropes without much difficulty and brought the ship safely into the hangar, while the crew of *Città* looked on.

After the landing it was sometimes difficult to obtain the information we required because we became dependent on *Città* for communications with Malmgreen and Nobile. As often as routines would permit, one of our radio operators listened to the messages and small talk exchanged between *Città* and Rome, but little useful information was obtained. We wrote off as malicious gossip a message saying that Nobile was superstitious and would not consider starting his polar flight on a Friday.

An excursion was planned to explore North Land,[22] and on our qualified advice *Italia* took off on May 11. A local snowstorm caused icing on the canvas, and Nobile decided to return to base before they had left Spitzbergen. A second attempt, starting on May 15, was largely successful. The weather en route was excellent, but minor snowstorms over North Land were bothersome; the ship reached 79° N and 100° E, when Nobile decided to return without having accomplished much exploration.

Summarizing the results of these two flights and of the *Norge* expedition in 1926, one might well have come to the conclusion that dirigibles can be of little use in the Arctic unless weather conditions are ideal. Clearly, icing and gusty winds were limiting factors of formidable dimensions. Nevertheless, the flight just completed, which had lasted sixty-nine hours, was taken to hold out promise of success in reaching their main publicity objective, the North Pole.

On May 22, 1928, Nobile was ready for his big adventure, and the ship took off at 4:30 the next morning. If everything went well *Italia* would be over the Pole on Friday, May 24. The weather situation seemed favorable provided that they sailed westward to Greenland, then to the Pole, and returned the same way. For two good meteorological reasons the area to the north of Spitzbergen was considered hazardous. Our experience indicated that it would be wise not only to recommend one route but also to stress the hazards of the alternate routes they might wish to choose.

The ship took off, flew due north, and soon ran into snow and gusty winds. *Italia* signaled for advice, and Tromsø repeated the earlier warning against flying straight toward the Pole. The ship then began to ease westward, eventually reaching the northeast corner of Greenland, where it changed course and headed for the Pole. From now on we had hardly any contact with *Italia*, but our "bugging operator" listened to the exchanges between *Italia* and *Città* and to the chitchat between *Città* and Rome. It appeared that the weather was fine. The Pole was reached, flags, crosses, and souvenirs dropped, emotions displayed: viva the Pope! viva Mussolini! viva Nobile! viva everybody!—except, perhaps, Captain Romagna. The ceremonies lasted almost two hours. Then began the fateful homeward journey.

For reasons that have never become clear to me, *Italia* chose to return via the area against which we had warned so strongly. We were not even informed of the change of plan. For a long time we heard nothing from Malmgreen. Neither did our "bugging" of the communications between *Città* and *Italia* and *Città* and Rome yield tangible information.

Then, on the morning of May 25, our radio operator brought me a message from *Italia* to *Città,* which said, "We are a hundred miles north of Moppen Island and fighting bravely against fog and strong winds; expect to be in Kings Bay in about two hours." The message was in Italian and slightly garbled; two hours seemed an unbelievably short interval. I told the operator to keep listening, but nothing more was heard. After several hours, we concluded that *Italia* was no longer in the air.

From information that became available later we learned that *Italia* had crashed on the ice at about 10:30 A.M., well to the northeast of Moppen Island, about 180 miles from Kings Bay.[23] The longitude of the crash was about 29°E. While we had recommended return via northeast Greenland, we had warned specifically against approaching Spitzbergen to the east of 10°E.

In the crash, the ship with the gas bags became separated from the gondola; the balloon rose, drifting northward, carrying six crew members in her belly. Nothing has been heard of them since. The remainder of the crew—nine men, including Nobile, Malmgreen, engineer Natale Cecione, first officer Adalberto Mariano, and second officer Filippo Zappi—survived the crash. Both Cecione and Nobile had broken legs. Nobile also had a broken right arm and was severely handicapped by several other injuries. The rest of the crew had escaped with minor injuries. Nobile's pet dog, Titina, was as fit as ever.

The downed crew built a camp out of the gondola and the wreckage, which later became known as the Red Tent. Supplies, which had been scattered over a large area, were collected, a spare radio transmitter and receiver were found to be in working order, and hopes were high that Captain Romagna would find some way of helping.

On May 26, the day after the crash, as prominent men in Oslo were gathering for a dinner to honor Sir Hubert Wilkins and Captain Carl Ben Eielson, rumors began to circulate that *Italia* had crashed or emergency-landed on the ice. Amundsen was present, and although no one could read his mind, everybody knew of his quarrels with Nobile and their extension to include Mussolini. At a suitable moment during the dinner, Amundsen let a friend announce that he was ready to organize and lead a coordinated effort to rescue Nobile and his men.

Amundsen's many admirers saw in this announcement an elder statesman offering a helping hand to an old enemy: The White Eagle of the North had risen to the occasion! But Mussolini saw things differently. The Norwegian prime minister's proposal for a coordinated rescue effort under Amundsen's leadership was turned down with a crude snub: Italy could manage her own affairs.

Without any coordination, and in some cases with scant competence, a number of rescue expeditions set out. Subsequently, when a would-be rescuer got into trouble, other expeditions were set in motion. Eventually, France, Italy, Finland, Norway, Sweden, and the USSR, as well as a number of private individuals, became involved. Coordination of effort was not only lacking; it did not even seem to be desired. We in Tromsø could not escape the impression that a tough competition was in progress. The business of rescuing the *Italia* crew had developed into a race among the various expeditions to be the first to bring back Nobile. An atmosphere of prestige and newsworthiness was clearly in evidence.

In the meantime, Nobile's camp kept calling for help, but *Città* was not listening, and the transmitter was not strong enough to be heard beyond Spitzbergen. Romagna had assumed that *Italia* had crashed on the ice and that no radio equipment could have survived the impact. Having made this assumption, he saw no point in listening for calls. Instead the *Città* radio station became occupied with chitchat exchanges and personal messages to family and friends in Italy.

In an effort to seek assistance, Malmgreen, Mariano, and Zappi set out on foot [from the Red Tent], hoping to reach Spitzbergen before the ice disintegrated. The complexities increased further when Captain Gennaro Sora was accused of "committing an act of serious insubordination." Sora, an able Alpine soldier, disobeyed Captain Romagna and, with two companions, set out on foot [from the *Città*] to find Nobile's camp. From our listening to *Città,* we obtained the impression that Sora intended to go eastward across Spitzbergen. We then informed *Città* that in our opinion *Italia* was most likely to be found in the Moppen Island area. Whether this message influenced Sora's plan is not known, but his excursion became directed northeastward rather than eastward. Sora's party did not reach Nobile's camp, however, and Malmgreen's party failed to reach Spitzbergen. Eventually search parties had to be sent out to find the searchers, and, as the second-generation searchers got into trouble, the rescue efforts became diversified beyond all expectations.

The first glimmer of hope came to Nobile and his companions on June 18. A Norwegian seaplane with two highly experienced explorers, Captain Riiser-Larsen and Lieutenant Lutzsow-Holm, was seen to fly almost over the Red Tent, but the camp could not be identified from the air. However, the flight convinced Nobile and his men that active searching was in progress.

The day before, on June 17, Major Umberto Maddelena, of the Italian Air Force, had reached Vadsø, hoping to continue to Kings Bay the next morning. Since the weather situation was favorable, we advised an early start. The flight was uneventful, and the next day, June 19, Maddelena took off from Kings Bay. He came within a few miles of the Red Tent, but failed to identify the camp. Improved markings were clearly needed.

At about this time events took place in rapid succession. On June 18 a Finnish expedition, headed by Lieutenant Sarko, was waiting in Tromsø, hoping to fly to Kings Bay the next morning. Sarko's seaplane was heavily loaded and would need some wind to become airborne. Next to Sarko's plane was an aircraft belonging to a Swedish expedition under the command of Captain Einer Lundberg, a daredevil and soldier of fortune. They, too, wished to reach Kings Bay on June 19. Two pilots, Sarko and Lieutenant Carlson, a Swedish officer with whom I had contact, had agreed to fly their planes together. The preliminary forecast was for suitable flying conditions, but since the general situation might not be very stable, we agreed that they should call for a final briefing at 4:45 A.M. A third expedition, an Italian seaplane under the command of Major Penzo, was waiting to take off, but apparently Italian policy was "to go it alone." Penzo called on me, got a favorable forecast, and departed without seeking other contacts.

Roald Amundsen, who had not succeeded in his original plan to head a coordinated rescue effort, had approached French authorities and had received a prompt response. The French had developed a new type of flying boat; one of them, the *Latham*, captained by Commander René Guilbaud and carrying a French crew of six, had been placed at Amundsen's disposal. On June 18, the *Latham* arrived at the crowded airport in Tromsø. In his later years Amundsen hardly ever put in an appearance without being accompanied by an aide. On this occasion his companion was his trusted friend Lief Dietrichson, a collaborator in the 1925 expedition that had almost reached the North Pole. Everybody felt that Dietrichson's participation added much competence to the enterprise.

Like Carlson and Sarko, Amundsen planned to take off early on June 19, but there was a difference: Amundsen's destination was a closely guarded secret. Though facilities and service were available at Kings Bay, the secrecy seemed to suggest that he might be planning something unorthodox. Perhaps he might have some idea of Nobile's whereabouts and chance a direct flight to the Red Tent. Many expected Amundsen to succeed where others had failed. While in Tromsø, Amundsen was staying with his close friend, apothecary Fritz Zappfe. All his meteorological wishes came to us via Zappfe and Krogness. Not knowing the *Latham*'s destination, all I could do was to reproduce the forecast I had given to the Finns and the Swedes.

Carlson and Sarko called rather earlier than arranged. Carlson, who spoke for both, said that they had hoped that Amundsen—as the grand old man—would have offered them help and advice. In particular, they had hoped that the three planes could fly together and help one another in case of need. Optimistically I thought no difficulty could possibly stand in the way of this excellent proposal. I suggested that they ready their planes and prepare to meet Amundsen where a boat would be taking him to the *Latham*. In the meantime, I would ask Krogness to ask Zappfe to suggest to Amundsen that he take these young pilots under his wing. Krogness liked the plan and phoned Zappfe, but returned a disappointed man. In substance, Amundsen had said that he could not be bothered with all these so-and-sos who knew nothing about the Arctic.

Very soon after the telephone call, the *Latham* departed alone (Fig. 3.2). The morning was sunny and calm. Carlson's Swedish plane rose and turned northward, as the *Latham* had done earlier. The Finnish plane could not take off without wind, so Sarko, with much regret, had to ship his plane and impedimenta to Kings Bay. About three weeks later he had the satisfaction of rescuing one of the would-be rescuers.

Although Zappfe did not know the *Latham*'s destination, he had told Krogness that Amundsen would like to receive reports on lake ice on Bear Island, in case an intermediate landing should become desirable. In response we arranged for our observer to collect information and be prepared for a direct call from the *Latham*. A few minutes before 11:00 A.M. the *Latham* called. Our operator responded that he would call them back in a few minutes, the reason being that he was about to take his turn in a tightly scheduled exchange of international weather reports. About ten minutes later Bear Island called the *Latham*—several times—and received no answer. Nothing was ever heard from the

Fig. 3.2: *Latham* airboat, Amundsen's ill-fated rescue plane, prior to taking off from Tromsø on June 19, 1928.

Latham again. It had not been seen or heard by our observers nor by any of the halibut fishermen that I interviewed upon their return from the Bear Island banks. After a long period of northerly winds one of the *Latham*'s floats drifted ashore somewhere on the north coast of Lapland; it was evident that the impact had been a violent one.

For a while the *Latham* disaster focused the public's attention on Amundsen rather than Nobile. An extensive search was made by several expeditions along the ice edge, even as far away as Greenland; there was always the chance that the *Latham* was still afloat. A French expedition, commanded by Admiral Heer, set out to search for the *Latham*. Heer had some data on how different objects drift when exposed to wind and ocean currents. Our calculations, which could not be very accurate, indicated that the float had drifted mainly with the wind and had probably started its independent journey somewhere near Bear Island. The weather had been fine. The most likely cause of the disaster was a mechanical failure. The *Latham* was then considered lost.

In the meantime, the Finns, Italians, and Norwegians continued their search for Nobile's camp and the Malmgreen and Sora groups. In addition, the old but still serviceable Soviet icebreaker *Krassin* plodded her way into the ice. On June 23, Lundberg, with his companion Lieutenant Schyberg, landed at the Red Tent and brought Nobile and his dog to Kings Bay. When they returned to fetch the much-injured Cecione, the plane crashed; the number of men in need of rescue had not been reduced.

Meanwhile, the *Krassin* arrived on the scene [and located Zappi and Mariano]. Malmgreen was not there; Zappi was in good shape and was reported to be dressed in some of the clothing Malmgreen had carried; Mariano, lying in icy water, was alive but badly in need of care, including amputation of a gangrenous foot. It was explained that Malmgreen had died of exhaustion long before and that it was his wish to be left behind. The *Krassin* also spotted the sprightly Sora group at a distance, but decided to attend to them later; finding the Red Tent had first priority. In the meantime, the Sora group was rescued by Finnish and Swedish planes. The *Krassin* found the Red Tent and brought all safely to Kings Bay, where the operation on Mariano was successfully carried out by Italian and Soviet doctors.

Captain Romagna, under orders from Rome, stoked his engines and returned home via Tromsø, where he treated my colleague, Sigurd Evjen, and me to a cup of tea. On her way south, *Città* visited a few Norwegian ports and was received with icy chill. The fact that Nobile had let himself (and his pet dog) be rescued first was considered unbecoming of a leader, particularly since Cecione was in greater need of care. Worse still were the ugly rumors centered on Malmgreen's disappearance. Unsubstantiated accusations of cannibalism received wide circulation. The loss of a national hero like Amundsen made it easier for uninformed people to heap suspicion and disgrace on Nobile and his lieutenants. Few considered the injustices done to such brave men as the Italian aviators Maddelena and Penzo, the insubordinate Captain Sora, and many of those in the Red Tent. The six men who disappeared with the balloon were mostly forgotten. Mussolini's ambitions, Nobile's dreams, and the obstructions of many ended in disaster without honor. Perhaps naïvely, I felt that the purity of the Arctic, which I had admired so much, had been soiled.

During the hectic days that followed *Latham*'s crash, newspapers in Oslo, Stockholm, Copenhagen, and Berlin were frequently on the telephone trying to squeeze information out of Evjen and myself. On one occasion, the line to Copenhagen was bad. When I asked for assistance at some intermediate station, a Swedish lady offered to serve as an intermediary. At long last, when the conversation came to an end, the lady said, "I believe I have heard your voice before." When I protested, she asked, "Didn't you buy some ptarmigan the winter before last?" We

Fig. 3.3: Petterssen on holiday near Tromsø with his sister in 1928.

chatted for a while and found it pleasant to recall the commotion that the ptarmigan transaction had caused.

For me, 1928 was a year of great change. The Geophysical Institute, understaffed and poorly financed, had become so burdened with routine forecasting and administration of arctic stations that there was hardly any opportunity for research and study. In the middle of the year the Institute went through a sort of binary fission: The meteorology part became a regional forecast center with responsibility for the arctic work, and the geophysical part, which had dwindled over the years, was replaced by a modern auroral observatory. I did find time, however, for a brief holiday (Fig. 3.3).

Much as I loved the Arctic I did not feel able to part from scientific meteorology. Fortunately, a vacancy had developed in Jack Bjerknes's group, and, before the year ended, I had returned to Bergen—with undying memories of the Northland.

CHAPTER 4

Bergen and Beyond

Modesty is an ornament, yet
people get on better without it.
—German proverb

Much had happened in Bergen since I left in 1924. Vilhelm Bjerknes, Bergeron, and Solberg had settled down in Oslo. And Jack, though still head of the Bergen weather center, was spending much of his time providing meteorological substance for a major undertaking headed by his father: the writing of a text on the physics and mechanics of fluid motion.

Having rejoined the Bergen group, this time on a more or less permanent basis, I reverted to the prediction problem, which had fascinated me so much in 1924. As I saw it, the problem was to develop methods for computing the movement and rate of development of storms. It soon became clear that, without upper-air observations, nothing useful could be done by starting out from the laws of physics, for all parts of the atmosphere interact, but the weather charts at that time showed only the conditions at sea level. Nevertheless, it proved possible to approach the problem on a different and far more limited basis. What I did was develop a series of simple mathematical expressions for the velocity, acceleration, and rate of development of weather fronts and storm centers, without asking *why* and *wherefore*. I had to be satisfied with just, *it is so*.

The Bjerknes influence was strong, and my Bergen colleagues were skeptical: Any "method" that was not rooted in the laws of physics was hardly a method at all. True! But what little I had done was useful, and it was the only thing that could be done at that time. It was only twenty years later, when upper-air data and electronic computers had become available, that practical forecasting could be treated as a physical problem. It took another decade before these new methods began to make inroads into the daily routines. On occasion it has been said that I had had "the courage to consider practical weather forecasting as a mathematical problem" and that I was one of the early contributors to the "*mathematisation de la prevision du temps*," [mathematicization of weather forecasting]. Though such statements may reflect some historical truth, the mathematical techniques now in use are far more advanced and reflect simplifications of the laws of physics.

Though analysis supplemented by simple computations had become routine in Bergen as early as 1931, I had no time to write a scientific paper on my findings. At this time my workload, which was already heavy, increased further when Jack resigned as head of the Bergen Center,[24] and I was chosen to succeed him. Though this promotion was both pleasing and encouraging, the freedom that Jack had enjoyed was not part of my inheritance. The economic depression in Norway was heavy; the routines in the weather service were increasing, and budgets had stagnated in spite of rising costs.

In the meantime, foreign visitors kept coming to Bergen. Some planned only short visits, hoping to learn a simple trick and return home to glory. Others, aware of the immense complexity of the problems, realized that much more needed to be done. They were, nevertheless, impressed with what we had to offer. Most of these visitors stayed several months.

One of the early visitors was Commander F.W. Reichelderfer, then in the U.S. Navy, who later became chief of the U.S. Weather Bureau; another was Andrew Thompson, who later became director of the Canadian Meteorological Service. Both occupied their high positions for a number of years and did much to modernize meteorological services in North America. A third—and very interesting—visitor was Father Depperman, S.J., who came all the way from Manila to become acquainted with the Bergen methods.

Like many of his colleagues in the Jesuit Order, Depperman placed more emphasis on practical work than on preaching. In the Far East, where typhoons cause so much damage on land as well as at sea, Jesuits had established a few observatories equipped with radio, and performed a very useful service by issuing warnings of the movement of typhoons and other storms. Many of these fathers had acquired extraordinary skill as intuitive forecasters; a few of them were highly competent scientists.

Father Depperman had been a secretary before he "went into religion." His shorthand and typing speeds were enormous; he seemed able to listen, talk, and type, all at the same time. Having seen me computing the speed and direction of movement of storm centers, he felt he ought to become acquainted with the theory underlying my method. One morning he walked into my office with his portable typewriter:

"Now let me hear your theory and I'll take it down," was his brisk request. Though I was unable to respond in the manner he expected, I supplied him later with a skeletal outline that gradually grew into a major piece of work, published by the Norwegian Academy of Sciences.[25] Father Depperman always expressed himself in positive terms. When he had finished the rough typing, he burst out: "Some consider weather forecasting a hopeless task; others say it is a skill that must be acquired; but now, we have *a method*!" Had he said that now we have something that may be capable of development, he would have been closer to the truth. Though my computations were useful, judgment based on experience remained an essential ingredient.

Even a storm may serve a useful purpose—unless it arrives unannounced. On the morning of February 1, 1933, an unusual storm development began near Iceland. It was not a regular Bjerknes cyclone; rather, it was a slow tightening of "the channels through which the air must blow": the narrower the channel, the faster the wind. Here was a case where my formula could give some guidance. There was much fishing going on at this time, and early warnings were much in demand. Soon gale warnings were issued for the following day, covering a coastal strip 400 miles long and centered on Bergen. But fishing was not the only problem. As a result of the economic depression, a number of unemployed ships were at anchor, and the Bergen harbor is particularly vulnerable to winds from the north-northwest. There was also a fire problem. Bergen, with its large number of wooden houses, had to take extra precautions when strong winds were expected. Fearing more than gale winds, I alerted the fire service and the harbor authorities.

Work to secure the ships started immediately and was intensified when warnings of storm winds were issued. Early on February 2 I went all out: *The storm winds will swing through northwest and increase to hurricane force late in the evening*. This warning was repeated in every broadcast during the remainder of the day. This, perhaps, is the only occasion on which winds of hurricane force have been predicted in Europe.

Winds from the dreaded north-northwest direction struck Bergen about midnight. The moorings held until about 5:00 A.M., when a few of the outermost ships began to drag anchor. For a while it was feared that these ships would play havoc with those closer to land. Fortunately, the wind veered further, and the Bergen harbor became less exposed; a narrow escape from a major disaster—in Bergen as well as in other harbors along the west coast.

The northwesterly winds had been extremely gusty and much damage had resulted. The reports soon made it clear that all damage had been inflicted on land installations; the fishermen, the harbor authorities, and all concerned with the sea had acted on the warnings and taken all possible precautions. Accounts of the storm, the warnings, and the damage filled front-page columns in many newspapers; some of them even reached as far as Oslo. Soon, an unofficial whisper told me that if I should request a small increase in my budget the application would be favorably considered. The response was favorable—and the increase was small. However, it was possible to believe that the tide had begun to turn.

Reichelderfer's visit had signaled an awakening interest in weather forecasting in the U.S. Navy. This interest gathered momentum in 1934 when Rear Admiral E.J. King[26] was placed in charge of the Bureau of Aeronautics. Commander (Pete) Hale was then sent on a five-month visit to Bergen. His primary mission was to prepare himself for a two-year tour as instructor of meteorology at the U.S. Naval Postgraduate School. As I learned later, Pete had instructions to pay brief visits also to other major meteorological centers in Europe; he was to return with a proposal for inviting a European meteorologist to give a series of lectures and laboratory instructions to naval meteorological officers.

Early in 1935, much to my surprise, I received an invitation from the Bureau of Aeronautics to give a two-month course at the Naval Air Station at Norfolk, Virginia, to be followed by a similar course at San Diego, California.[27] This invitation was followed by many others, but I could not accept them all. Having finished my work for the navy, I spent four months as a visiting professor at California Institute of Technology, Pasadena, and three months equally divided between the U.S. Weather Bureau and the Canadian Meteorological Service, with brief visits to a few universities. My ten-month leave of absence from Norway was fully packed with work.

I had prepared well for my courses and, fortunately, the U.S. Navy proved a model of efficiency. All services, such as typing, proofreading, drafting, and duplicating, were of the highest order. The Norfolk Naval Base was a very quiet place. My twelve students were divided into task forces. Every morning my lecture and demonstrations from the previous day were available in printed form. Within a month of my departure, the Bureau of Aeronautics had prepared an edited version of the course material—a neat book with a preface written by Admiral King. Though the book was intended for in-service use only, it received a wide unofficial circulation. Five years later it was published in a much expanded form.[28] By that time the United States was preparing for war. The sales of the book in the United States, as well as abroad, exceeded all expectations.

However, we must return to 1935 and my early impressions of America. As we approached New York late one evening, with the city lights as a distant background, I saw two noteworthy structures. One was the Statue of Liberty, shining her torch toward the Old World; the other was a huge light panel flashing a message to us newcomers. Perhaps naïvely, I associated the one structure with the other, wondering what inspiration might emerge. Straining my eyes, I could make out a message in three parts: WRIGLEY HERE, WRIGLEY THERE, WRIGLEY EVERYWHERE—a giant advertisement for the famous chewing gum. On the spur of the moment I found my naïveté rather amusing, but in later years, as I learned to know America better, I read something symbolic into what I had seen, a reminder of two traits so typical of America: idealism and commercialism. Expanding on such thoughts, I have sometimes said that the American mind is greater than its European counterpart: greater in good as well as in evil, greater in its love of political freedom, and more steeped in political corruption. In America the amplitude is very much larger. The phrase "The biggest in the world," which some Europeans find annoying, has substance behind it, though the distinction between *bigness* and *greatness* is not always clear.

I was immediately impressed with American hospitality. We disembarked on a Sunday morning, and I was due to leave for Norfolk on the night train. During the voyage, two of my table companions had planned my first Sunday in New York. Mr. A took care of me in the morning. After a charming lunch with his family, Mr. B took over. In an open car and with a sunny breeze we toured New York, drove along the Hudson River, visited the theater in Rockefeller Plaza, and had dinner in my host's club; then I was driven to the railway station.

The hospitality I had enjoyed so much in New York was equally impressive on the naval base and elsewhere in Virginia. Though the lavishness might vary according to circumstances, the friendly informality was general, and I felt at home wherever I went. I soon found the average Virginian to have a strong leaning toward the bourgeois side: proud of their history, neighbor-conscious, and mindful of respectability. (I was once told that in respectable Virginian families, the ladies hang their underwear on the clothesline covered by pillowcases.) Church attendance was widespread and the churches seemed anxious to cultivate a club atmosphere, with but little emphasis on the solemn observances so typical of the sparsely attended churches in the Scandinavian countries. Was Christianity on its way out? If so, it would certainly linger longer among the colored population.

However, Virginia is one thing and California another. Even in the aftermath of the Depression, life in California was carefree and often lavish. Soon after my arrival in Pasadena I was taken sightseeing in Los Angeles and the surrounding suburbs—textbook examples of Thorstein Veblen's *Theory of the Leisure Class*, except that conspicuous waste, rather than true leisure, was the dominant trait. My lasting impression of what I saw was the newness of it all; sightseeing had but little to do with visiting old structures. Our first stop was at the famous Suicide Bridge, with its tall span over a river that is dry most of the year. Here, wealthy men who had lost their fortunes during the Depression had come to end their troubles by jumping over the rail. As the traffic increased, police were stationed on the bridge to regulate conditions.

My knowledge of California, Arizona, and adjoining parts of Mexico was insufficient to appreciate much of what I heard and saw, and I have never been good at pretending. However, soon something happened that stirred my Scandinavian imagination. I was taken to lunch in the Trocadero, the famous place where Greta Garbo had snubbed Jean Harlow. It was explained to me that the snubbing reflected Greta's ignorance about the appropriate way of spelling Jean's surname.

The day was sunny, cool, and pleasant. We visited a number of other places of historical interest, ending the tour on the edge of the desert at a tiny church built long ago by a small English community. About 1920 the church had been deconsecrated and later turned into a candlelit restaurant with excellent food and service, but *positively no liquor*.

An hour or two after I had settled in at California Institute of Technology (commonly known as Caltech), a secretary phoned to say that the president wished to see me. The president was Robert A. Millikan, of cosmic rays fame, a Nobel laureate in physics (1923). The Americans, unlike many Europeans (including myself), have a wonderful way of meeting people. Millikan was no exception. It was evident, however, that something was weighing on his mind. He soon came to the point: He had made a bet with an army sergeant on a point of physics, and had lost a case of beer. Worse still, the uncouth sergeant had teased him about his Nobel prize. Ugh! Millikan had just returned from an army base in San Antonio, Texas, where he had sent up high-level balloons to measure the intensity of cosmic rays with a minimum of atmospheric interference. Like many scientists at that time he had taken it for granted that the upper winds would be from the west "because of the earth's rotation." However, for reasons that were none too clear, the sergeant held a different view and won the bet. I did the best I could to explain that the winds were "normally" from the west, but, in summer, when the "heat belt" shifts northward, easterly winds spread northward at high levels over the subtropics. Millikan did not seem convinced, but when I offered to direct him to a book by Vilhelm Bjerknes, all arguments came to an end—though his mystification seemed to linger.

Under the dynamic leadership of Millikan, Caltech had risen to the forefront of learned institutions. Meteorology, which was a new activity on the campus, was part of the Department of Aeronautics, headed by Theodore von Kármán, a Hungarian Jew who had risen to great fame while working in Germany. Another famous man, with whom I had close contact, was Beno Gutenberg, a world-famous geophysicist, a direct descendent of *the* Gutenberg and the owner of one of the first bibles ever printed. On occasion, Albert Einstein would visit his friends at Caltech. His aura of fame, learning, and wisdom was unmistakable.

Gutenberg had left Germany quite early, while von Kármán and others had emigrated when Hitler's anti-Jewish policies began to become clear. While the Nazis felt they could do without Jewish scientists, von Kármán was an exception. The *Luftwaffe* had to be expanded. Göring pressed the Secretary of Education and put pressure on von Kármán to return to his aeronautical chair in Aix-la-Chapelle. Von Kármán either ignored these messages or answered them in a light and evasive vein. Von Kármán was a typical Jew [in appearance], particularly when seen in profile; he was known to have had communistic leanings; thus he was as vulnerable as anyone could be in the Third Reich. Eventually he was outright ordered to return. To bait the hook it was stated that his outstanding ability in aeronautics would entitle him to a privileged position. On this occasion von Kármán pretended helplessness, took the message to Millikan, and asked for advice. When, after some considerable discussion, Millikan asked in despair, "What *are* you going to do?" Von Kármán, answered, "I'll send him my profile"—the word *profile* being frequently used in technical aeronautical language. The end result of all these exchanges was that an undying joke spread over the campus.

Millikan was not only a great scientist but also an astute administrator. While most private universities had obtained their endowments through last wills and testaments, Millikan had managed to collect large sums from living individuals who, without much loss of time, could tour the campus and read their names in large gilt letters on the walls of research laboratories. Why wait until it is too late? I was told that during the Depression, Caltech had been able to help one or two donors who had parted with too much. Generosity, to be meaningful, must come from strength.

Millikan and von Kármán had become convinced that the expansion in aviation, typical of the thirties, would place increasing demands on meteorological services. They decided to try to establish a school for training practical forecasters. The MIT school, headed by Rossby, was thought to be too academic. They soon found their way to Dr. Irving P. Krick, one of Gutenberg's geology students, who had had some experience in airline forecasting. Apparently, the purpose of inviting me to Caltech was to assist in attracting students to the meteorology school that was then in the making. As we shall see in later chapters, meteorological activities on the Caltech campus did not enjoy a long life. Nevertheless, Krick was a delightful person, and through him and von Kármán I got to know a number of charming people, most of whom were connected with the motion picture industry, the press, and aeronautical activities.[29]

While at Caltech I received a message from Oslo informing me that a committee had been appointed to make a study of civil aviation in the United States and to prepare technical reports that could serve as a basis for the planning of civil aviation in Norway. For reasons of economy it was considered important that, as far as possible, facilities and services for civil and military purposes be combined. The members of the committee were Bernt Balchen (civil aviation), Bjarne Øen (military aviation), Odd Nansen (buildings and installations), and myself (weather services). Having foreseen the need for meteorological planning, I had completed a study before I received the message. When the committee began to work, all I had to do was to make minor adjustments to accommodate the wishes of the other members. The reports were well received. As the need for international coordination became pressing, Norway was well prepared to play an active part.[30]

I returned to Norway in April 1936. As a result of the 1933 publication of my paper on forecasting methods,[31] I soon became involved in the work of the International Meteorological Organization (IMO), first in matters directly related to forecasting, then in applications to shipping and aviation, and, ultimately, in the exploration of the upper atmosphere. During the years before the war, international conferences took me to Copenhagen, de Bilt, Utrecht, Paris, Lisbon, Berlin, Munich, Salzburg, and Vienna. However, to attend conferences is one thing; to prepare for them and work at and between sessions is something else. For a person like myself, who was concerned not only with the administration but also with the science of meteorology and its applications in many walks of life, the conferences left but little time for extracurricular activities. Nevertheless, wherever I went, I was fortunate enough to meet a large number of interesting people.

During a lengthy aeronautical conference in Paris in 1937, I had the good fortune to become acquainted with the Janneres family. M. de Janneres, a well-known musician and a brother of the famous architect le Corbusier, lived somewhere on the edge of Bois de St. Clou in a house designed by his brother. Though I had read a good deal about le Corbusier's creations, it was heartwarming to see this charming dwelling, particularly as compared with the pictures I had seen of his many provocative extravagances. In architecture, as in art, fine miniatures are rarely signed.

A conference of exceptional interest was held in Berlin only two months before the outbreak of the Hitler War. The atmosphere was tense, and one could discern happy as well as worried faces. "Hitler will ruin the *Vaterland*," said an elevator operator, when we were not overheard; "Hitler will gain for us everything without war," said a fellow scientist; "If the choice is between bread and honor, I choose honor" said a meteorological administrator; and "history—the verdict of history? *We* shall write the history!" said a *Bierstube* [beer hall] boaster. Far more frightening: "*Wir lieben vereint; Wir hassen vereint. . . .*" ["We love united; We hate united. . . ."] was sung by a group of small children while being paraded by their teacher. Civilization gone mad!

In itself the meteorology of the meeting was not very important. The conference was the umpteenth session of the much-neglected Maritime Commission of the IMO. The Commission's president, Dr. van Everdingen, a distinguished Dutchman, had tired of German intrigues, resigned, and sent one of his men to handle the business until a new president could be elected.

A few years earlier the German meteorological service had been reorganized in a most un-German manner. The service, whose structure was already complex, was to be led by a headless quadrumvirate of directors: one for science, one for climatology, one for weather services, and one for maritime affairs. The four directors neither loved nor hated such organization. What little cohesion there was came from the circumstance that all knew that Göring might frown even on his favorite, Dr. Habermehl, who was in charge of weather services, including all *Luftwaffe* business.

Though formally just one of the four, Habermehl had powers and channels not available to the others. Habermehl was a moderate Nazi, in favor of the *Lebensraum* and *Herrenvolk* ideas,[32] but rather opposed to the violent anti-Jewish movement. "What is the point of destroying so much valuable property?" he asked. Habermehl was proud of the service he headed. One evening, in strict confidence, he told us that Germany had available 2,700 meteorologists, most of whom had doctoral degrees. Though the definition of a meteorologist was none too strict, it was clear that their state of preparedness was impressive. At this time Great Britain had about one hundred meteorologists, France eighty, the U.S. Air Force about thirty, and Norway fewer than twenty.

The maritime interests of Germany were presided over by Rear Admiral Spiess, who was often looked upon as a freshwater admiral. Another prominent member of the conference was Captain L.G. Garbett, brother of the Archbishop of York and head of the Meteorological Service of the Royal Navy. One evening Spiess took Garbett and myself for a stroll down Wilhelmstrasse. It was about 8:00 P.M. when we passed the Chancellery where Hitler was entertaining an important guest—Prince Paul of Yugoslavia. I was surprised to see that only two guards were visible, and I was overwhelmed when Spiess, dressed as a civilian, stopped one of them and said, "I am Admiral Spiess of the *Deutsche Seewarte*, and here are Captain Garbett of the Royal Navy and Dr. Petterssen of the Norwegian weather service. Could we just look inside?" There was nothing to it; the answer was "Yes." The sight was impressive: a huge bare hall of reddish marble, beautiful in a way, but conveying a frightening impression of massive might. The corridor to the left seemed uninteresting, so we turned right, in the direction whence Wagernian music came. As we approached the door to the banquet hall, the receptionist said in a quiet tone, "*Nicht weiter, bitte*" ["No further, please"].

After the stroll we returned to an important evening session of our commission. The last item on the agenda was the election of a new president. By all customary rules, involving rank, seniority, geographical distribution, etc., Spiess was the obvious choice. However, his election might conceivably be taken to indicate increased prestige for one of the four German directors. Also, as the conference progressed, intrigues had been set in motion, and one or two of the delegates had obtained new instructions from home; a mild anti-Nazi breeze was in the air—mild and late. When the election came up, it so happened that Spiess was the only German present and could not readily vote for himself. It was now past 2:00 A.M., and the atmosphere had become childish in the extreme. I had come to the conference as a junior member and had with me a plan for weather service for ships on the high seas. With later expansions, the plan has remained in force; one must assume that it has served a useful purpose [see Chapter 23]. Be this as it may, there can be but little doubt that intrigues, converging from different sources, played an important part in events: I was elected to succeed van Everdingen.

As the new president of the Maritime Commission, I had to appear in Berlin nine days later to present the session's report to the International Meteorological Committee. Because I had accepted an academic position in the United States, I had to go to Bergen to attend to pressing business and then return to Berlin to present my report. During my second visit, Göring invited us to a reception in *Das Haus der Flieger*, a rather palatial club building for *Luftwaffe* officers and other persons associated with German aviation. About forty guests were lined up in a U-shaped pattern, all with their backs to the wall. At the appointed time it was announced that unforeseen events would prevent Göring from attending; he would be represented by two of his high generals, Fritz and Stumpf. In came F. and S., clicked heels, and greeted each guest with the indoor Nazi salute. The "big animals" were closest to the entrance, and each responded with a limp salute. I knew I could not do it. When the turn came to me I offered my hand—Norwegian fashion—and said in English, "How do you do, Sir?" Much to my surprise the two generals responded in like manner and with perfect grace. I watched the rest of the line; only one other person did not respond with the Nazi salute. Mr. E. Gold, deputy director of the British Meteorological Office and president of the Commission of Weather Services, greeted the generals as I had done. Apparently Gold, too, had acted spontaneously. I have always felt proud of my misbehavior in *Das Haus der Flieger.*

CHAPTER 5

New Horizons

Go west, young man!
—John Soule[33]

During my visit to America I had had several offers of permanent employment, all well paid but somewhat varied in quality. One of them seemed highly rewarding, especially when translated into work opportunities. The U.S. Weather Bureau had long been in the doldrums. Although "fancy European ideas" did cross the Atlantic, the Bureau's "considered opinion" had always indicated that things were different in North America. Early in 1935, a new chief of the bureau had to be found. On the recommendation of his scientific advisers, Isaiah Bowman, Robert A. Millikan, and K.T. Compton, President Roosevelt appointed Dr. Willis Gregg, an able administrator with a progressive outlook. Gregg had been advised to search for a deputy who could take charge of a program of research, education, and scientific services. After about four weeks of work in the Bureau, I was asked if I would consider an offer. Gregg went out of his way to assure me that acceptance would be well received by his staff. Since we were both due to attend an international meeting in Salzburg in the autumn of 1936, I promised to let him have a firm answer by then.

For a number of reasons, including personal ones, I had to decline the offer. I thought then that no comparable opportunity would ever come my way again. Most unfortunately, Dr. Gregg died a year later, and Roosevelt's advisers had to search again. This time, the choice for the top job was between Reichelderfer and Rossby, both of whom were familiar with the so-called Bergen methods, Rossby far more than Reichelderfer. The outcome of these considerations was truly Solomonic. Reichelderfer, a skilled and cautious administrator, was to succeed Gregg; Rossby, who was recognized not only for his scientific skill but also for his flair for innovation, was persuaded to accept the position of deputy chief, with responsibility for the scientific advancement of the Bureau. Everyone knew that Rossby, with his vision and drive, would soon make his influence felt in wider spheres. These decisions having been reached, Rossby spent several months in Bergen; after some time, he inquired whether I would consider an offer to succeed him at MIT. The terms were very favorable and I had no difficulty in making up my mind.

I had hardly settled in at MIT before Rossby came to see me. Before I had even had an opportunity to explore the campus, I had agreed to give a six-week course in weather analysis and forecasting for promising Weather Bureau employees. Rossby, I found, had his own peculiar way of transacting business. I soon discovered that all arrangements, including my dean's concurrence, had been made before I had been approached. Rossby had simply taken my willingness for granted. My very helpful dean had been unable to imagine that Rossby had come to him before he had obtained my enthusiastic concurrence. Rossby, however, felt he was on firm ground with his explanation: "We need your course very badly; I have never seen you miss any opportunity to do a useful job."

The Weather Bureau had space and staff available at Newark Airport. Thither I went, almost empty-handed, to give a series of lectures and laboratory instruction. The students were enthusiastic and the arrangements excellent; as usual, Rossby had acted with a speed that no bureaucrat could match.

After Newark I went to Washington to renew contacts. I was especially anxious to report what I had seen and heard in Berlin only a few weeks before. No excitement was aroused when I reported what Dr. Habermehl had told me about their manpower preparedness; 2,700 meteorologists seemed just like boastful thinking; an atmosphere of smugness was clearly in evidence. No one had considered that under shifting wartime conditions manpower cannot be deployed economically; to be prepared for war means to have *at least* enough. In Washington, as in Berlin, many members of the "bureaucratic intelligentsia" believed that Hitler's appetite was limited and that he would achieve his goals without war. Most convincingly, Czechoslovakia was so very close to Russia. I returned to MIT thinking of Washington as a gigantic anthill in which each little ant runs back and forth carrying a tiny twig or a little piece of stubble, without knowing why and wherefore. Regardless of plans and the quality of management, some anthills grow bigger and bigger. As far as is known, none has ever gone bankrupt. What I had seen in Berlin had convinced me that war was not far off, and I was not impressed with what I had heard in Washington.

Back on campus I found that Rossby also had talents as a publicity agent. I found that I had been well advertised as a hard worker and an efficient manager. My dean, Dr. Moorland, went out of his way to be considerate and helpful. As a matter of routine, the budgets had been approved before I arrived, but if I needed more all I had to do was say so. He even offered extra funds to provide me with a personal assistant; both he and the president were anxious that I not waste time on work that could be handled by others. I felt that I was playing with Aladdin's lamp.

These many luxuries enabled me to concentrate on the completion of the manuscript of my textbook, *Weather Analysis and Forecasting* (1940), a piece of work that I had been struggling with, on and off, since 1936. By Christmas the manuscript was ready, and the publisher rushed the manufacture of the book. The publisher, who thought that war was on our doorstep, went on to convince me that there was an urgent need for a companion book, a nonmathematical one, for the training of subprofessional personnel. Within a few months my *Introduction to Meteorology* (1941) was ready for processing. Taking stock of the situation, I found that, compared with the lean years in Norway, my productivity had increased tenfold, all due to the responsive management of MIT.

Though I had considered the outbreak of war in Europe to be imminent, I was, nevertheless, stunned when it came. Never before had I felt so attached to Europe as I did on this fateful September 1. The falling bastions in Central Europe, the twilight policies of Chamberlain and Halifax, the flimsy leadership of Daladier,[34] the scared military leaders of France and Great Britain, the hedging of the pope, and, on top of it all, the widespread isolationism in the United States, held out but little hope for survival of what we, proudly and loosely, call western civilization. Old and supposedly strong structures were crumbling. If there were strong political leaders, they were silent—they had been so for many years. Men of reason, Europe's pride in earlier centuries, had no place in the Hitler era of Wagnerian music, brute force, "degenerate art," and moral *Mitläuferei* [running-along-with]. Europe was aflame, and the firewatchers had not yet had their morning coffee. Yet I felt deeply attached to Europe—the Europe of Erasmus, John Locke, Goethe, Tolstoy, and Zola.

In early 1940, Captain Arthur (Merri) Mereweather, then chief of the Army Air Corps, had about thirty "weather officers," all pilots whose ambition it was to advance, not in meteorology, but in command positions. And war or no war, thirty officers seemed not an unreasonable number for a captain to handle. I told Merri that the *Luftwaffe* had 2,700 meteorologists, all civilians put in uniform for the duration and given rank to correspond to the assigned responsibility. This time I met with a response. Merri asked me to write a report to the War Department with an estimate of the "meteorological strength" of Germany, Italy, France, and Great Britain. Not long thereafter, I received a telegram, signed by General Arnold, asking whether I could accommodate special courses for "meteorology cadets." The MIT response was immediate and enthusiastic.

At about this time, Rossby had become interested in training programs. Working with the air force and the navy, he set in motion a nationwide effort. The first MIT program got under way in June 1940; soon other universities followed. Each year the number of students increased. Everyone agreed that we needed *more*, but no one tried to decide *how many*. The program continued to snowball long after a balance between supply and demand had been reached. But long before then, Rossby had left the Weather Bureau, and I had left MIT.

I found academic life in America rather different from what I had seen in Europe. In America, meteorology had never been among the most attractive fields of learning. Many students were job hunters rather than seekers of knowledge. "What kind of job will I get if I take these courses?" and "What pay can I expect?" were questions I had to try to answer when I interviewed students. At times it was not easy to decide who was interviewing and who was responding. The students' approach was always straightforward and without pretense. Many were out to acquire a skill rather than basic understanding, but they knew what they were after, and they worked hard to achieve what they sought.

The "pedestal professor" of Europe had no place at MIT where the student-professor relationship was close, informal, and productive. My door was always open. The blackboard in my office was used frequently, and so were the paper napkins in a modest restaurant across the street where we gathered for lunch—keeping the cost as close to thirty-five cents as we could manage. Some of my more venturesome students even joined my Sunday skiing school in the hills north of Boston.

Though few of my students were scholastically inclined, their inventiveness was always impressive. Time and again I handed out subjects for master's theses and found that acceptable answers were produced without much work. On other occasions, inventiveness and hard work combined to give excellent results. Mr. X, who had no strong inclination to theory, was asked to produce a descriptive study of the distribution of frozen subsoil in the eastern part of the United States. I had assumed that the basic data were available somewhere in Washington and that he could provide a useful study without an undue amount of work. Finding that the basic data did not exist, Mr. X set out on his own. He bought 150 postcards, composed a questionnaire and mailed the cards to that many undertakers between Lake Michigan and Maine. The cost of a grave depends very much on the depth of the frozen layer, and efficient undertakers know very well how to calculate their profits. It was as simple as that. Mr. X's grade was a well-deserved A.

Though MIT was an institute of technology, it was also in the forefront of basic learning and academic teaching. Here the professors were engaged in *doing* as well as *teaching*. Since industry was moving into the Boston area to be close to the crucibles of knowledge, it was MIT policy to leave the professors free to accept consultant work unless they felt it interfered with their regular duties. The management had seen to it that there were few rules and regulations to observe. At the time I joined the MIT faculty, the directors of the S. Morgan Smith Company of York, Pennsylvania, had decided to explore the possibilities of giant wind turbines as an additional source of power. To harness the wind on a large scale required knowledge of certain aspects of wind regimes, which we did not then possess. It was understood that a considerable amount of research had to be done before a practical solution could be found.

The idea of developing giant wind turbines, rather than conventional mills with battery storage, had originated with Palmer Cosslett Putnam, a Harvard geologist turned engineer and promoter, who became the leader of the project. Putnam invited me to head a meteorological team to explore wind regimes and develop site selection criteria for blocks of turbines in Vermont, where a utility company had indicated interest.

The work branched out in three directions. Professor Wilbur, my colleague at MIT, designed the tower, a complicated structure that could not only carry the turbine but also withstand stresses caused by storm winds and loads of ice. Von Kármán designed the wind-catcher, a hun-

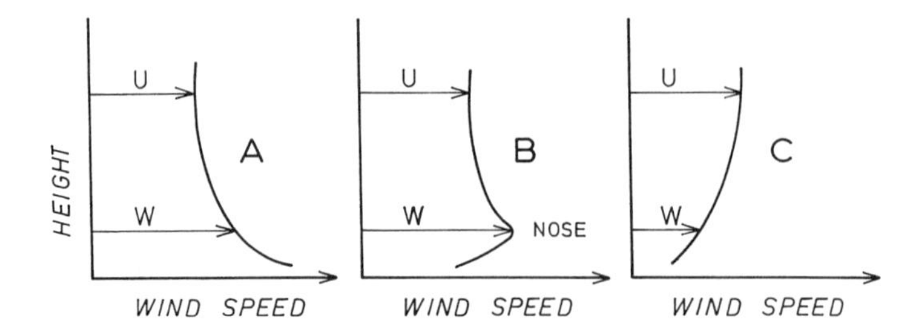

Fig. 5.1: Types of wind profiles over a mountain ridge. The "wind-nose" is of particular importance to wind turbines.

dred-foot turbine blade with many sophisticated gadgets. The meteorology—the wind regimes, icing storms, and damaging gusts—became my domain. This turned out to be the least explored area and it soon became the crux of the project. Putnam, with his driving energy, was a typical promoter, and believed that money and manpower would solve all problems. I soon found myself in charge of a think tank of four subconsultants (of whom Rossby was one) and a large field program directed by Professor Karl O. Lange, a German expert on instruments and measurements. The potential buyer of the power was brought into the consultations at the very beginning, and the project was pressed forward with undue speed. As a result, I soon became too heavily engaged in consultant work. Since some of my not-so-well-off students earned their living by part-time work on the turbine project, there was no easy retreat. Eventually, I made a resolution never again to become involved in industrial consulting work.

The basic problem of how to harness the energy of the winds on a large scale was in itself intriguing, though the commercial possibilities were obviously marginal. At a certain distance above water or flat land one may speak of an undisturbed wind. However, when the air streams over a ridge, the wind regime becomes disturbed, so the local wind above the ridge may be weaker or stronger than the undisturbed wind at the same level. The deviation of the wind field depends on the shape of the ridge and on the amount of friction, or the roughness of the ridge. If the shape of the ridge were ideal and friction were absent, the wind profile above the ridge would be of the type shown in diagram A in Fig. 5.1. However, friction is always present and will reduce the speed close to the ridge. If the ridge is of suitable shape and the surface relatively

Ridge Factor W/U	0.6	0.8	1.0	1.2	1.4
Power (U = 1.0)	0.2	0.5	1.0	1.7	2.0

Fig 5.2: Ridge factors and power of the wind.

smooth, the wind profile will be about as shown in diagram B: There will be a "wind-nose" at some distance above the crest (Fig. 5.1). Finally, if the shape of the ridge is awkward and friction strong, a profile like C will be found. In nature one finds all kinds of variations between B and C; only theoreticians are interested in A.

As far as wind turbines are concerned, the wind-nose is extremely important. If U is the undisturbed wind and W the actual wind, the ratio W/U is called the *ridge factor*: It is a measure of how much the wind has been increased or decreased above the ridge. In nature one finds ridge factors varying from 0.5 to about 1.5. However, other conditions are also important. The energy of the wind is proportional to the square of the wind speed, that is, W^2. The power corresponding to this energy is $W \times W^2$ or W^3. Figure 5.2 shows how the power of the wind varies with the ridge factor. The power is expressed as a percentage of the power corresponding to the undisturbed wind U.

Although the energy of the winds is enormous, it is of the low-grade variety, in the sense that its concentration is very low and it is not easy to collect it in neat bundles. Unless we could find sites with ridge factors well above unity, the turbines could not compete with conventional methods of production.

Though the concept of the ridge factor is very simple, its measurement is rather difficult. The atmosphere is an erratic system, and a long series of measurements along tall masts is required to obtain reliable values. My early estimates had indicated that it should be possible to find ridges in New England with a factor approaching 1.2. Though a few such places were found after lengthy series of measurements, we failed to develop methods for selection of sites without building masts and measuring over many months. The results of our measurements showed that most ridges were substandard. Thus, to find one good site, one had to measure at several places. Without much simpler procedures, site selection would add substantially to the initial cost.

Another obstacle in the way of economic production was the fact that New England, on the border between moist Atlantic and bitter arctic air masses, is an El Dorado of severe icing storms. To safeguard blocks of turbines against icing damage would add much to the initial cost as well as to the operating expenses.

Though our findings did not point toward profitable operations, the directors of the S. Morgan Smith Company desired to satisfy their curiosity and ours; they decided to build a full-size test turbine on Grandpa's Knob in Vermont, where the ridge factor was favorable. It turned out that the turbine could produce electricity at a cost of about $0.06 per kilowatt-hour which, at that time, was found to be too costly. The company, having spent well over a million dollars on the project, decided against continuation and, very generously, placed all patents and technical information in the public domain.

From the beginning I had had doubts as to the wisdom of building turbines in New England where icing storms are frequent and the wind direction and speed are quite variable. The ideal is, of course, to have a strong steady wind across the ridge. In the course of the study I developed a world map of power regimes and emphasized the advantages to be found in the trade-wind belt and the monsoon regions, where icing is no problem, the winds are exceptionally steady, and electricity is quite expensive. Though the use of wind turbines may be limited to special regions, they may, in the aggregate, amount to a considerable business market and a blessing for many poor people. In the final analysis, wind energy is nothing but realized solar energy; it is already there, and it is not associated with pollution.

CHAPTER 6

A Visit by Friends

En art il s'agit d'être intéressant.
[In art it is necessary to be interesting.]
—Jules Laforgue

In early July 1941 my good friend Nordahl Grieg (Fig. 6.1), a famous Norwegian poet, playwright, and novelist, and a close relative of Edvard Grieg, came up to Boston to see me. To my great delight the famous Ibsen actress Gerd Egede Nissen had recently become Mrs. Grieg.

I first met Nordahl in 1937 when he suddenly walked into my office in Bergen, apologized profusely for disturbing me, but seemed at a loss to explain why he had come. Nordahl had a naïve respect for science and believed that deep thinking was going on all the time. As it happened, I was working on a trivial budget matter and was quite glad to be interrupted; also, his smile was completely disarming. We soon felt at home with one another. Both of us had a strong liking for the *Nordland* and an enthusiastic conversation followed. Then he came to the business of his call. He had been invited to write a short story for a new edition of *Norsk Lesebok* [*Norwegian Reading Book*], a collection of outstanding pieces of prose and poetry used in elementary schools to introduce children to good literature. Although most of it is composed of selections from the classics, some of the short stories have a slant toward reality. (Of what I read in my schooldays I still remember, most vividly, "Torje Wigen"— an epic poem by Ibsen—and a short story about whaling in which Svend Foyn, the inventor of the harpoon, was the hero.) Nordahl had accepted the invitation and planned to write a story in which weather forecasting, especially storm warnings for the Lofoten fisheries, would play an important part. Nordahl had read Stewart's *Storm*, and someone had told him that the equation that the forecaster looked up—and did not use—had been taken from my doctoral dissertation.[35] Nordahl had come to ask if I would help with advice on factual and technical matters related to storm warnings and their use by fishermen. My thoughts went back to my early childhood, my grandfather's many encounters with the raging elements, and my mother's "human scale" of the force of the wind. Needless to say, I was pleased that Nordahl had

Fig. 6.1: Nordahl Grieg (1902–1943),
friend, author, patriot.

come to see me. During the following months we met on several occasions, but after Lord Runciman's mission to Prague in 1938, we lost interest in the *Lesebok*. With Europe clearly heading for a major upheaval, it was difficult for me, and far more so for Nordahl, to concentrate on a long-term cultural project.

Nordahl and I never argued about Hitler or Chamberlain, but we disagreed rather strongly about Stalin. With an introduction from no less a person than Fridtjof Nansen, Nordahl had visited China and spent two years in the Soviet Union. He had interviewed Chiang Kai Shek and Stalin, and had been to rather boisterous vodka parties with Gorky, Molotov, Voroshilov, Budenny, and others.[36] He wrote against Franco during the Spanish Civil War.

Before the outbreak of the Hitler War, Nordahl had strong communist leanings. For my part, I considered tyranny in any form to be the overwhelming evil. I could see little difference among Hitler, Stalin, Mussolini, and Franco—except in terms of stupidity which, unfortunately, was not limited to the dictatorships. After the outbreak of the Hitler War, Nordahl changed his position. By the time we met in Boston he was simply a humanist and a patriot; political ideologies had been replaced by a deep sense of duty—a duty to promote justice, human rights, and human dignity and, above all, to work for the liberation of Norway. Later, when I came to London, I was disappointed to see how few Norwegians understood Nordahl. In this world, a new reputation seems to be the most difficult thing to acquire. Fortunately, several members of the Norwegian government in exile, including secretary of foreign affairs Trygve Lie,[37] fully appreciated Nordahl's position and had no doubt about his ability to stimulate interest in the cause of Norway.

In the early years of the war many American newspapers ascribed the sudden collapse of the Polish army; the occupation of Denmark, Norway, Belgium, and the Netherlands; and the defeat of France to widespread treason. In Norway, about three percent of the electorate belonged to a Nazi-type political party. Since its leader was the renegade army officer Quisling, his name became used (as if it were the present participle of an onomatope) as a synonym for treason.[38] The true state of affairs was well understood in Europe, but some American journalists went so far as to have Norway appear as a hotbed of treason. To counter this completely unjust and harmful publicity, Trygve Lie and other members of the government arranged for Gerd and Nordahl to go on an unofficial goodwill mission to the United States and Canada.

While living in Norway, I had hardly ever missed an opportunity of seeing Gerd playing Ibsen. Her *Hedda Gabler* still ranks above anything I have seen in or out of Norway. But I had never met her in person before she and Nordahl arrived in Boston. Gerd, who had started her career as a child actress, had known the royal family since she was a young girl. These connections had become firmer after the spring of 1940, when the king and government escaped Hitler's assault on Norway. As a result of introductions at the highest level, Gerd and Nordahl had been received by the president. Gerd had indeed carried a personal letter from King Haakon to President Roosevelt.

Nordahl and Gerd were free to do anything they thought useful to improve the image of Norway, and they traveled widely to visit centers of Scandinavian interest. Their most pressing problem, however, was one in motion picture production. Before the outbreak of war in Europe, Nordahl had commenced work on a script with the title *Greater Wars*.[39] As the war situation hardened, he revised the script several times and included events related to the Quisling treason. Nordahl tried to bare the roots of the everlasting conflict between good and evil and trace them back to an imbalance between our endeavors in science and technology on the one hand, and in the humanities on the other. He tried to show that, without humanism, scientists may gradually, and sometimes unwittingly, become the tools of Nazism. On this basis, he tried to explain how the Quisling treason and similar events had grown out of recent cultural trends. Nordahl's thesis was that people must unite against wars and join to gain mastery of the forces of nature to improve the lot of humankind. In line with this he had chosen as the chief character of *Greater Wars* a meteorologist, working in storm-swept north Norway, trying to do his part by studying the nature of storms and improving weather forecasting. The final scene had a man and a woman standing in a graveyard on the stormy coast of north Norway reading data on the dead: They were mostly women and children—the storms of the past had left so many widows.

King Haakon had read the synopsis of *Greater Wars*, gave his enthusiastic support, and offered to serve as protector. President Roosevelt, too, read the synopsis, let his admiration become known, and asked the writer and playwright, Robert Sherwood, to do the monitoring. Encouragement came in generous amounts from many influential people, including Sigrid Undset, Mrs. Borden Harriman, Wendell Wilkie, Mayor La Guardia,[40] and a wealthy New York financier who preferred to remain anonymous.

Nordahl was working hard on another script, based on the life and times of Edvard Grieg. The Oscar-winning actor Paul Muni was especially interested in this latter film. Nordahl was greatly overworked and the oppressive heat and moisture added to his discomfort. He was definitely in a bad state and needed rest more than exertion. Only when Gerd was with him did he appear to be calm and composed. The best I could do for them was to drive them up to a charming old inn on the top of Mount Mansfield in northern Vermont, about 4,000 feet above sea level. It was a Saturday and I could stay over until Sunday afternoon. After supper Gerd, having discovered how much I adored Ibsen's women (particularly as compared with the bamboo-legged spinstery creations of Bernard Shaw), acted *Hedda Gabler* for me while she or Nordahl supplemented with sentences spoken by Løvborg and Brack. I was deeply stirred by the depth of her acting. I returned the next weekend, and one evening Gerd acted parts of *Greater Wars*. But for the guns in Europe, life was at its highest heights.

Some months later, when we met in London, Nordahl was kind enough to say that he had been stimulated in his writing of *Greater Wars* by our early conversations in Norway. Although Nordahl did not know the meaning of the word flatter, there can be but little substance in his remark. In our conversations I never went very deeply into my personal problem: that of a pacifist–scientist in a technological culture where humanism has become little but an ornament. I cannot even claim to have been consistent in my personal brand of pacifism: My two daughters, Eileen and Liv, were trapped in Norway. At times I felt outright revengeful and wished I could join in fighting against the Nazis. Although my days were completely filled with useful work, I was often impatient, for I felt it might take a long time for public opinion in America to swing away from the various shades of isolationism that existed, particularly in the Midwest. No democracy can fight a war successfully without the support of an overwhelming majority of the people, and, in the summer of 1941, such a majority simply did not exist.

After Gerd's acting of *Greater Wars*, I told her and Nordahl that I found it unsatisfactory not to be directly involved in the war against Hitler. About six weeks later I received, via the Norwegian Embassy in Washington, an urgent message from London. But that belongs to a later chapter.

What became of the film projects? Both got stranded. The main obstacles that stood in the way of the Grieg film were due to Paul Muni. He found the script too Norwegian and would act *Edvard Grieg* only if the text were completely revised to meet Hollywood style. Nordahl found the demands unacceptable. However, the really important project was *Greater Wars*, and here several hurdles had to be overcome. One obstacle was a law passed in January 1941 prohibiting the making and showing of certain types of war films. Although this was a hobble to Roosevelt's wheelchair, American politicians are normally versatile enough to find a way if the will is there—and there were powerful wills at work. A major difficulty was the general background. While Sikorski, prime minister of the Polish government in exile, was queuing with manuscripts for propaganda films, the Jewish component of the motion picture industry was being criticized by the isolationists for feeding the public with warmongering films.

Nordahl probably created difficulties for himself by being rather inflexible; his film was art, not propaganda. He did not realize that the film industry, like all other business enterprises, has no soul. The leaders may smile at profits and blush at deficit, but human dignity is not in their books. When Roosevelt, who was trying hard to help, asked if he thought *Greater Wars* would pay, Nordahl seemed astounded at the thought that idealism had to make a profit. His film was not written for that purpose; the liberation of Norway was at stake.

Though Nordahl was unbending, he was not so in the everyday meaning of the word; he was a poet and his actions were governed by idealism. He was untiring in visiting the Norwegian army, navy, and air force units, hoping to inspire them and the people at home to believe in the liberation of Norway and the coming of better days. But Nordahl felt, very strongly, that he could not write or talk unless he himself lived through the dangers that the men—in barracks, ships, aircraft, and hospitals—had to face.

Norway lost a cultural hero in December 1943. Until the summer of that year, my friend Nordahl Grieg had joined in many ventures of the Norwegian army and navy. He had taken part in several flights over Norway and sailed in Atlantic and Arctic convoys. Now, a feeling that he must also experience the dangers of bombing raids began to take hold of him. Berlin was his preferred target. His arguments were strong: "To know and describe, and to tell people at home, is my work," and "to lighten their burden is my duty." He was met by an equally strong opposition. Because of the dangers involved, many friends tried to talk him out of his plan of going on bombing raids. Knowing that Nordahl actually sought danger, I tried to persuade him by saying that he, being untrained, big, and bulky, would endanger the safety of the crew should anything go wrong. When he came to Dunstable [see chapter 9] to see me I took him to the nearby deer park at Whipsnade and, in very pleasant surroundings, begged him not to add to the load of the RAF crews. But Nordahl had to write, and he could only write what he lived.

For a long while everybody in authority refused him permission to join in a raid. Then, to the surprise of many, the decision was suddenly reversed. I did not know which raid Nordahl would be permitted to join.

But it so happened that I helped to prepare the forecasts for the raid on December 2, 1943. I sat up, at Dunstable, all night preparing ten-minute drift corrections and wind forecasts for the bombing force. In that raid, one of the RAF's biggest, an Australian bomber was shot down over west Berlin. Later, a golden chain and heart, a gift from Gerd, identified Nordahl. A later raid obliterated all traces of his grave.

Like Edvard Grieg, Nordahl's patriotism knew no bounds, and his love of Norway was matched with a deep sense of the quality of life and a desire to give. Long before the outbreak of the war, Nordahl wrote:

> Nought have I given, but prove me,
> All that is mine demand;
> Youth and its fire—come, claim them,
> Hallow them thine, my land!
> Thus to be my loving tribute
> Not in dead words expressed,
> But to take my life as a garment
> Cloaking thy naked breast.[41]

Sometimes I felt that he considered death for a noble cause as the highest form of fulfillment, something that he not only desired but actively sought.

CHAPTER 7

To England via Newfoundland

It irked him to be here; he could not rest.
—Matthew Arnold

Late in August 1941, I received an urgent message from my old acquaintance Admiral Riiser-Larsen, commander in chief of the Norwegian Air Forces, asking me to come to England to provide weather forecasts for a squadron with frequent missions involving Norway.[42] When the war broke out Norway, with barely three million people, was the world's third-largest shipping nation and was leading the world in tankers. At the time of the Nazi invasion, in April 1940, the overwhelming part of the merchant fleet was on the seven seas and became available to the Allied war effort. With generous income from shipping and by using their substantial borrowing capacity, the Norwegian government tried to build up forces to accelerate the day of liberation.

An overriding consideration was to strengthen the links between the administrative and the political powers: the government in exile and the underground movement at home. Strengthening these complex connections, encouraging opposition to the Nazis, arranging for balanced refugee traffic, integrating policies and plans, and stimulating cohesion were problems of paramount importance. It was in these areas that Riiser-Larsen's force could do useful work far out of proportion to its size. Clandestine airdrops and landings within Norway provided much-needed supplies and support for the underground movement. The main link, however, was with the Stockholm airlift and various surface channels, mostly through the wilderness that separates Norway from Sweden. For political as well as other reasons, the traffic between Sweden and England had to be augmented at a rapid rate. Our pilots, flying over enemy-held territory, needed suitable cloud cover rather than clear skies.

The visit by Gerd and Nordahl had whetted my desire for active participation in the war, and, on the face of it, nothing could be more satisfying for me than to be useful in work related to the home front. However, few things are as simple as they appear at first glance. Normally, work at the squadron level is limited to adjusting basic forecasts, conducting briefings, and, when the need arises, local liaison work and dignified salesmanship. I knew Riiser-Larsen from the prewar years and thought he had a weakness for the grand gesture; perhaps he had not taken sufficiently into account the fact that the basic forecasts had to be provided by the British Meteorological Office, and that I might not be particularly useful at the local level. Would I have the staff and facilities needed? Had suitable arrangements been made with the British?

There were other complications. I was just about to wind up my second academic year at MIT. There had been twenty-four students in my department during my first year, and ninety-six in the second. Between the two years, in the summer break, we had conducted a special ninety-day meteorology course for instructors of air-force personnel. I was now preparing to receive 155 students for the coming year. I had managed to add a number of young instructors to the staff. Though they were exceptionally bright, they needed general guidance and the support that long experience adds to book knowledge. Although America was still neutral, our program aimed at preparing for war. As a result, our teaching was heavily concentrated on areas that would be immediately useful. Of foremost importance was weather forecasting, and none of my instructors had much experience in this field.

Now that I had an opportunity to become involved in the war in Europe I felt torn between loyalty to my staff and students and my patriotism and duty to Norway. There would have been no question of choice had I not been much in doubt about my usefulness in the work that Riiser-Larsen had to offer. Dean Moorland, who had been so helpful when I first came to MIT and had helped me adapt to campus policies, took me to see our president, Dr. K. T. Compton, a real Solomon, loved and admired for his wisdom, firmness, and kindness. Having heard all sides of the problem, Compton said he would decline an application for leave of absence and back the refusal with a statement of how useful my present work was for the overall war effort. His statement would prevent, or disarm, any comment that might reflect unfavorably on my patriotism and sense of duty toward Norway. I felt that I had come to a fork in a road and had been assisted to a wise decision along factual rather than sentimental lines. However, sentiment does not always yield to wisdom, and my restlessness did not subside.

In the meantime, Riiser-Larsen persevered. A second telegram, expressed in very strong terms, left me with but little choice. Compton and Moorland agreed with me. I was offered indefinite leave of absence and compensation for any financial loss that might result from the recall to service for the Norwegian government. Such generosity, I felt, could be found only in the United States.

But complications are apt to cluster. In the previous academic year, I had introduced in the forecasting laboratory a special course in ocean analysis, or, rather, weather analysis and forecasting for ocean areas. Particularly during the colder part of the year, both the frequency and the intensity of cyclonic storms are far greater over the northern oceans than over the adjacent continents. If the United States went to war, the demands for weather forecasts for the ocean areas would increase enormously while, for security reasons, the number of transmitted reports from ships would be greatly reduced. There was, therefore, an obvious need for a course of this type. One of the strong points of the so-called Bjerknes School of airmass and polar front analysis is that its techniques facilitate the combination of rather widely scattered observations into coherent representations of the broad features of the anatomy of storms, thus providing a closer point of departure for predictions. My exercises in ocean analysis had been well received by the naval officers who had attended my classes; in early 1941, when Admiral King took command of the Atlantic Fleet, all efforts to improve the state of preparedness became greatly accelerated. The first sign was a large reduction in the number of typewriters and rubber stamps, and an unmistakable emphasis on techniques and training.

Admiral King had been one of the early promoters of meteorology in the U.S. Navy; his interest never waned as he advanced to high command. An officer with responsibility for readiness in the meteorological field had prevailed upon me to spend a few weeks at sea to repeat my ocean analysis course for some senior meteorological officers who, for good reasons, could not be spared for regular schooling. I had promised to squeeze in four to five weeks, beginning early in October; I felt that I had to honor this commitment before going to England.

I had excellent connections within the navy and all arrangements were made in first-class style. A large yacht, which had belonged to an automobile magnate, had been acquired by the navy to be used as a messenger ship. Its name was *Zircon*. This very comfortable ship was detailed to pick me up at Boston Naval Yard on October 5 and take me to Argentia, Newfoundland, where I was to be housed on the *Albermarle*, a brand-new aircraft tender commanded by Captain Mullinix, who was locally called "the Commodore." The *Albermarle* was to provide all facilities for my course, and my students were posted to her. It was a pleasant surprise to find that one of the officers had attended my courses at San Diego in 1935 and two were students during my first year at MIT.

The *Zircon* was captained by a lieutenant commander, obviously a professional sailor, though apparently not a regular naval officer. Had he been older, I would have thought that he had had his training at the time when iron men commanded wooden ships. In contrast, almost all of his officers were "one-stripers," tall and handsome college boys put in uniform, out to begin to learn the arts of seamanship. About two hours out of Boston we ran into foul weather and most of the youngsters turned seasick. I had been working unusually hard to clear out at MIT and to prepare for a new adventure, and I felt greatly in need of rest. So I turned in immediately after supper and, in spite of the noisy retching of seasick sailors, I dropped off immediately and slept for eleven hours straight. The next day the weather was tolerably good but few of the youngsters turned up for meals. The skipper volunteered some sarcastic remarks about modern youth, seasick sailors, and other things not to his liking. The *Zircon* was not a fast ship, and, for the first time in many years, I felt no need to hurry. The smell of sea air, the salty spray, and the wide horizons came to me as a major blessing—as nature's lovely gifts. God, in His wisdom, had made the oceans large.

One morning just before first light, as we were passing somewhere to the south of Halifax, an explosion awakened me. Something was obviously wrong; however, I was sleepy, and, since no alarm had been sounded, I did not feel obliged to get up. Then, suddenly, it dawned on me that since many of the officers were seasick, the captain might need more help than might emerge if his young officers tried to respond to an alarm. As I rushed to slip on my dressing gown, there was a second explosion, and a third as I ran up the ladder to the bridge. The captain was there, cool as an iceberg. In his colorful language he explained that we had encountered a Nazi submarine and, perhaps unwisely, dropped depth charges. Sonar observations had shown that the sub was quite a distance away. She and perhaps others were waiting for convoys out of Halifax and did not wish to attract attention by sinking our little tub. Anyway, the *Zircon* had no business looking for a fight; we could have saved our depth charges and the bangs. We went on our way. A message was sent to Argentia, received, and laughed at. At this time German submarines had not been observed west of mid-Atlantic. A sub? Out of the question! A whale, perhaps, but not a sub! However, the captain was nobody's fool. He had a good card up his sleeve: His instruments had recorded the sound of a propeller that was switched off after the second depth charge was dropped. There was no need to call in a consultant in marine biology.

The weather improved toward noon and the sea was quite pleasant. The captain decided to teach his men a lesson and ordered gunnery practice. I volunteered to help, weighting the balloons so that they would barely rise while floating in a moderate westerly breeze. To me the weather looked ideal for beginners to get good marks. The practice got underway and smoke puffs appeared over a wide area—but none near any of the balloons. The captain's face darkened, but he said nothing. As the exercise came to an end, he inspected the guns. The last shot in one of them had not been fired. The silence was broken by his loud command: "Shoot." The order was obeyed instantly and, without aiming, the gun was fired, hitting a balloon that was by now quite far away. In spite of the fine weather and nature's blessings, I decided to have a rest. I spent much of the remainder of the day in bed, reading Shelley's "Ode to the West Wind."

Soon after the *Zircon* encounter, packs of convoy hunters were observed to the south of Nova Scotia and also outside the St. Lawrence estuary. Worse still, the U.S. destroyer *Kearney* was torpedoed in the mid-Atlantic and barely managed to reach Iceland. The *Kearney* was captained by Commander Antony Dennis, a close friend, who spent six months with me in Bergen in 1938 to prepare for a teaching assignment at the Naval Postgraduate School. The submarine hunt was on; Roosevelt's shoot-at-sight order was in force; and the voices of isolationism were definitely growing fainter. Perhaps the tide was beginning to turn.

Life on the *Albermarle* was exceedingly pleasant. "The Commodore" proved to be a fine host, a real seaman, and a deep thinker, all in one person. On movie nights he invited me to dinner and, afterwards, on the scheduled minute, as we walked into the theater, all officers and men rose to stand like tin soldiers until we were seated; the same exercise was repeated at the end of the show. With my sloppy habits I found this parading somewhat uncomfortable, but I had to agree that the anatomy of discipline is a complex affair. It is easy to be critical of individual components and to underestimate their importance in the structure as a whole. From my observations of Captain Mullinix, I concluded that the best disciplinarian is one who has no need to show his displeasure.

The nearest town was St. John's, the easternmost and one of the oldest towns in North America. In many respects it reminded me of some of the towns along the Skagerrak coast of Norway where trade prospered and wealth accumulated during the era of sailing ships. The uplands seemed exceptionally bleak and uninviting. It was difficult for me to recognize any feature that could have inspired Leif Eriksson and Thorfinn Karlsefni to attempt colonization, and especially, to name this rocky land "Vinland the Good."[43]

One Saturday afternoon I was taken sightseeing in St. John's. Before leaving, one of the senior officers, who was in the habit of proffering gratuitous advice, told me that certain standards in St. John's had declined in recent years, notably since construction works got under way at Gander and Argentia. Although his tortuous discourse lacked specifics, I understood that nowadays the difference between hospitality and immorality was at low ebb. Although my companions were ready to call it off, I decided that, in my case, the risk was not unacceptable. I found the people of St. John's very charming, and I returned the following Saturday for more sightseeing. Again I was impressed by the similarity to certain nonexpanding towns along the coast of Norway where traditions and an atmosphere of seamanship still linger. People who depend on the sea for their livelihood are always very friendly.

While at Argentia I received a message through the British Embassy in Washington, saying that a seat was reserved for me on a Ferry Command flight leaving Gander on November 8, and that the equipment I needed for the flight would be brought from Washington by a returning British mission. Another message, from Riiser-Larsen, told me that the director of the Meteorological Office, Sir Nelson Johnson, had asked that I be loaned to the British Service. Arrangements would be discussed upon my arrival in London.

At Gander I took the opportunity to see how Patrick McTaggert-Cowan operated the weather service for Ferry Command, whose primary function was to bring American-made bombers to England. I had heard much, and well-deserved, praise of Dr. Mac's skill at running a fine service under very difficult conditions, and I heard more as the war progressed. In due course a plane arrived from Washington with three men whose destination was London. They brought with them my outfit: fur-lined boots, a double flying suit, a pair of enormous gloves, and a fur cap. Concentrating on seeing as much as possible of meteorology in operation, I paid little attention to my companions: an admiral in the Royal Navy, a high official in the Ministry of Aircraft Production, and a certain Professor Whitehead. Then, as Dr. Mac was giving his finishing touch to the en route forecast, it dawned on me that the elderly professor was *the* Whitehead [Alfred North Whitehead], the famous mathematician–philosopher. As has happened so often—to my disadvantage—I had been too absorbed in my work.

To avoid a so-called weather front, which stretched halfway across the Atlantic, we had to fly high, 15,000 to 18,000 feet much of the way. A strong tailwind in the border region between deep polar and tropical air masses brought us to Prestwick, Scotland, in record time, as I recall, nine hours and eight minutes. From the point of view of creature comfort it was a miserable flight. The bomb bay was drafty and let in air far below freezing; there were no seats except baggage and boxes of a variety of shapes; and the noise made conversation pointless. To get some sleep I stretched out on some packing cases and, most unwisely, released my oxygen mask. I woke up as we began the descent toward Prestwick. My legs and arms felt leaden, and my perception was not at its best. My companions had kept their masks on and they, including Whitehead who was then over eighty, were wide awake and keen as schoolboys: They had flown the Atlantic! Not feeling any kind of exaltation I reminded myself that Lindbergh, Balchen,[44] and others had done this almost fifteen years earlier, and much had happened since. Moreover, they were pilots and navigators, not just cargo as we had been. After landing, someone produced the wherewithal for a small celebration. Although I began to feel brighter, I found it difficult to mobilize much enthusiasm.

Whitehead was remarkably lively and stimulating beyond description. He had chosen to leave Harvard and return to England to work for the Ministry of Economic Warfare. Although my academic home was next door to his, we had never met, and I knew him only by reputation. Naturally he had not the faintest idea about my background, except what he had been told in Washington: I was a Norwegian meteorologist who had immigrated to the United States just before the war in Europe broke out, and I was now on my way to England to work with the Meteorological Office. He thought it laudable that I, a non-British person, would choose to leave a professorship at MIT for the reasons that had been explained to him. Since the whole ensemble of real reasons was rather difficult to explain, I simplified matters. With a considerable element of generalized truth, I said that if the British Empire, western civilization, and all the rest of it were to go to pieces, I should like to be there to see how it happened. In meteorology I expected to have a grandstand view, and admission was free. The admiral laughed heartily at this explanation. None of us was optimistic enough to hope that America would soon enter the war.

While waiting for a plane to take us to London, Whitehead and I drifted into a loose conversation about general trends. Although my association was with technology, I had come to feel that what we often, and with pride, call "western civilization" was becoming increasingly weighted toward materialism or, rather, toward the materialistic by-products of science and technology. We were, I thought, turning our backs on Socrates's dictum that what matters is the ordering of human affairs rather than the control of nature. I felt that we needed a technology matched with a basic philosophy—a new kind of religion. The conversation now took on a new quality. I suddenly felt like a youngster sitting at the master's feet. A deep sense of history came over me and I recalled, very vividly, one of Edvard Munch's murals in the Aula of Oslo University. I was reminded again of our conversation when, in 1951, I saw (and bought) Thomas Hart Benton's lithograph *Instruction*. Sitting there at Prestwick Airport, I felt as if I was listening to a modern Socrates, or a twentieth-century Erasmus: a philosopher who had sought and found an integrated view of what we loosely call life. I felt that I was listening to a logical and methodical soul. I was very sorry when a conscientious airport official interrupted us.

Whitehead was the most reasoned and reasonable person I have ever met. He would surely be a boundless source of inspiration to the group he was now about to join. But his advice on my petty personal problems did not reveal much of a practical grasp. He told me that London taxicabs were so cheap and readily available that it was hardly worthwhile using public transport. A month later when, as a matter of routine, I had to get from the Admiralty to South Kensington at 2:00 A.M., I found that I had to resort to the old-fashioned method of putting one foot in front of the other and, additionally, to do some map reading in the blackout.

Professor Whitehead gave me another piece of practical advice. All I knew about the blitz and the bombing was what I had read in American newspapers or heard on the radio. I wondered whether any of the few London hotels I remembered were still standing. When I asked where I could find lodgings near the center of London until I had sorted out my work program, Whitehead thought for quite awhile and then said, rather philosophically: "If I were you I would go to either the Ritz or the Savoy." Before I could formulate a supplementary question to indicate more clearly what I had in mind, he continued: "After all, the Ritz is slightly on the vulgar side. I should go to the Savoy." Our companions, the admiral and the official of the Ministry of Aircraft Production, seemed to concur. I believe it was the admiral who called an RAF sergeant to book a room for me at the Savoy. The risk of landing me in slightly vulgar circumstances had been neatly avoided.

We took off in the early afternoon and landed at Northolt, where a staff car was waiting to take us into London. My companions insisted that the Savoy be the first stop. I thanked them all for their many kindnesses, and we parted, expressing hopes that we would meet again. We never did.

Chapter 8

Loaned to the British

Each venture is a new beginning.
—T.S. Eliot, *East Coker*

The next morning I began a round of calls, starting at the Norwegian headquarters. Refugees had been trickling in from a variety of directions, and all had to come to London to get a fresh start. Being anxious to hear the latest news from Norway, I joined a small group waiting in the lobby. Here I got a fine bird's-eye view of the general situation. Many seamen who had sailed in foreign ships had left their ships, some without observing customary procedures. Most of them had joined in convoy work, but a few had insisted on being enrolled in the armed forces. Some brave men and women had set out in fishing smacks across the North Sea, but not all had managed to escape the *Luftwaffe*. The majority of the refugees found their way through woods and wilderness into Sweden. A few of them, if they qualified as high-priority cases, came out by the Stockholm airlift, but the bulk got stranded in Sweden, halfway between home and destinations:

Scenes pure though full of peril
You left—no turning back.
Your goal was arms and freedom—
Closed too that forward track.
You, in mid journey stranded,
With other thousands caught,
To halfway life seem fated,
Twixt what you lost and sought.[45]

Before the Nazis invaded the Soviet Union some refugees had left Sweden and traveled around the world, via Vladivostok, Tokyo, Shanghai, and Hong Kong, while others reached their destination via Bombay and similar ports. There is nothing like Norwegian patriotism if it can be combined with individualism. Most of the young men who came to North America went to "Little Norway," a training camp operated by the Norwegian Air Force on the outskirts of Toronto. Then they were shipped to England.

It was really touching to talk to these young men and to observe how eager they were to join in the fighting. I remember especially a young man who had come to London via Shanghai and Little Norway, and was now about to be posted to a Norwegian squadron. His home was on an island somewhere between Bergen and Trondheim. In the early twenties I had served as a sergeant in an infantry regiment that received conscripts from these islands. Among us "professionals" we used to say that what these islanders had in common with the ducks was that they both moved more easily on water than on land. I doubt that any of us could beat them at rowing, but to teach them drill was something else. It seemed to me to be something quite out of the ordinary that this young islander should choose to travel across Eurasia to get to England. And when I asked him how he hit on the idea he answered in an offhand manner: "Oh, I knew that if I only got to Shanghai, I could sail the rest of the way." His story proved to be somewhat simpler when told in toto. He and his elder brother had decided to escape the Nazis. Since they lacked connections with the underground men who operated the "North Sea Bus Company," they decided to go to Sweden. Somewhere in eastern Norway they became separated and only one of them could continue across the border. The Norwegian Embassy in Stockholm provided some money, and a man with shipping interests in the Far East took the youngster, who knew no foreign language, under his wing. At this time it mattered little which port, one could always expect to find a Norwegian flag in its harbor. So sailing was no problem.

All the refugees I met had exciting escape stories to tell. As a newcomer I must have seemed exceedingly dull. The encounter, on October 8, with a German submarine was past history, and no one could possibly be interested in Whitehead's philosophies.

My first business call was in the Norwegian Ministry of Defense where I happened to meet a high civilian official, an elderly retired colonel. The Ministry of Defense, like other Norwegian institutions, was trying to organize not only to meet customary requirements, but also to accommodate refugees according to their number, gifts, and skills—all unpredictable quantities. The colonel was very kind and offered to help in anything I might wish to do. He inquired about my travel expenses and was surprised to hear that I had had none. He was obviously short of work—very short. I thanked him for his kind offers and said I would return when and if problems should arise. I left him with a feeling of strange sympathy—a bureaucrat short of paperwork, reminiscent of the headless woman in the *Pickwick Papers* who sat with a bun in her hand and no mouth to put it in.

I next called on Riiser-Larsen and heard, to my great satisfaction, that he had become convinced that Sir Nelson Johnson had been right in pressing for my transfer to the British Meteorological Office. However, his own problems had to be solved also. In the first place, there was the Stockholm airlift, which had to be expanded and pressed to maximum efficiency. Second, as his force developed, there would be more frequent missions into our occupied homeland. Third, it might become necessary for the Allies to occupy northern Norway in order to secure the convoys to Russia. Though the Stockholm airlift was of immediate interest, we had to be prepared for expanded activities. He wanted me to consider how adequate services could be provided. His point was that forecasters with experience from Scandinavia and the Arctic were essential; did the British have such personnel?

I knew that one of our most outstanding forecasters, a certain Mr. H. Anda, who was known to have communistic leanings, had escaped the Nazi invasion and had arrived in England via Little Norway. We soon found that this exceedingly well-equipped man was serving as a sublieutenant on some obscure RAF station. Action was immediately taken to have him promoted and reposted. I knew also of another Norwegian meteorologist who had been stranded in Stockholm, so Riiser-Larsen ordered a request for priority on the airlift to be sent through channels. As to future arrangements, I undertook to encourage a number of Norwegian forecasters to escape to Sweden. Riiser-Larsen said that he would use his influence to secure seats for them on the Stockholm airlift.[46]

Riiser-Larsen was now very happy; it had taken us a little more than an hour and hardly any paper to solve his meteorological problems. He explained that I was now free to offer my services to the British Meteorological Office. However, there were certain strings attached: It had been decided that I was to be employed by the Norwegian government and loaned to the British. He said this in an apologetic tone, and I assured him that I saw nothing but advantages in the arrangement.

My next stop was at Victory House, Kingsway, to see Sir Nelson Johnson (Fig. 8.1). I was most cordially received and I obtained the impression that he was genuinely glad that I had come to work for them.

Fig. 8.1: Sir Nelson Johnson (1892–1954), head of the British Meteorological Office.

However, when it came time to discuss specifics, I found his language rather guarded. After some time I thought it wise to let him know of the loan arrangement, and to stress that I was interested only in work—technical work—directly associated with the war. I was not concerned about rank, status, and the like. As soon as the war ended I would return to Norway or the United States, depending on the situation at that time. Apparently, the elimination of problems related to pay, rank, and status took a load off his mind. More than three years later, when the war in Europe ended, Sir Nelson wrote me a warm letter of appreciation. Among other things, he wrote, "I shall never forget the way in which you came to put your services at our disposal."

After a few words about Norwegian meteorologists and their posting to selected stations, Sir Nelson asked about what kind of work I would like to do. In the field of technical work I could choose anything I liked. However, he had thought of three different activities in which I might have an interest.

First, with current plans for expansion of air operations, especially for the bombing of industrial centers in Germany, there was an urgent need for a vigorous attack on the problem of how to forecast the winds in the upper atmosphere. He explained, in considerable detail, what they were doing to develop upper-air equipment and measuring techniques for use in connection with aircraft and balloon soundings; measurements made by radar were under way, and instrument-carrying rockets were under consideration. In this work, which was monitored by Professor P. A. Sheppard, the Meteorological Office was undoubtedly leading the world. On the other hand, and unfortunately, they had done very little to explore the uses that could be made of this new information to predict the changes in the upper atmosphere, even over short time spans. In this area, Great Britain seemed to be lagging behind such nations as Germany, Norway, and the United States.

Second, there was the perennial problem of long-range forecasting. It had come to the fore at the outbreak of war and pressures were increasing: Planners always like to know everything. A large group had been established at forecasting headquarters, Dunstable, to test and evaluate different approaches. The group was presently directed by a statistician, and none of its members seemed to have any knowledge of forecasting. Would I?

In case neither of these areas should appeal to me, he had thought that I might like a roving assignment, to assist in the forecasting for major or especially important operations.

Without much hesitation I ruled out long-range forecasting. At best it seemed to be a problem for the Third World War. Moreover, I had no qualifications that would make me a useful partner in a statistical enterprise. Although the group at Dunstable had been at work for almost two years, nothing useful, or even promising, seemed to be within sight. Being a firm believer in liaison with reality and fearing the inertia that the group seemed to have acquired, it appeared reasonably certain that they would soon become a victim of Murphy's Law.[47] The roving assignment had some appealing aspects, but I feared that my skill at "Englishmanship" might not suffice. So without further exploration I told Sir Nelson that I had a strong preference for the upper-air work, though occasional assignment to special jobs might be accommodated.

I obtained the impression that my choice was as Sir Nelson and his deputy, Mr. E. Gold, had hoped it would be. I felt certain that future development in weather forecasting lay in the use of upper-air data. Since the Meteorological Office's effort in this field was next to negligible, I could get under way with my work with no inertia to overcome. The question of upper-air forecasting was obviously a pressing one. Two nights before, while I was crossing the Atlantic, an RAF force had suffered very heavy losses. Hitler had been scheduled to speak to the nation and, for political reasons, it was most desirable that he be bombed off the air. Strong head winds and icing on the homeward journey dispersed the force and many planes failed to return.

Before I left, Sir Nelson and Mr. Gold asked me where I lived; when I said "at the Savoy," both tried to hide their surprise. When I told them the circumstances that had led me to prefer the Savoy, rather than the Ritz, we all laughed, Mr. Gold immoderately. My first day in London had been very much to my liking.

Roving assignments came sooner than I had expected. Since certain arrangements had to be made before I could begin work at Dunstable, Sir Nelson suggested that I visit some bases in Scotland that supported missions of interest in Norway. In Edinburgh, my first stop, the hotel clerk alerted the police, and I had to prove to their satisfaction that I was not a German spy. The Meteorological Office had provided me with a bunch of cards and papers to which I had paid but little attention. When we discovered that I carried "the yellow pass," issued by the Foreign Office, all doubts disappeared.

Next night, at Rossyth, I slept in a bitterly cold barrack. At 7:00 A.M. I was awakened by a beautiful young lady—Wren[48]—with a large cup of steaming hot tea. In Scotland in winter, the Gulf Stream is wonderful, but a cup of hot tea is often more stimulating. Ten minutes later, when I felt like braving the elements (indoors), I discovered that my trousers had been taken from the room. Thirty minutes passed and I began to feel trapped; I was traveling light and had no alternative. Had this not happened at a base of the Royal Navy, I would have been seriously concerned. A little later the young lady reappeared with my trousers, neatly pressed. I was staying in the officers' quarters and the Wren had done nothing but perform a regular routine.

I was less fortunate at Scapa Flow. A howling storm was blowing and the rain "fell" in horizontal streaks. On two occasions I, being a civilian, was "bumped" by petty officers and had to ride in an open van while traveling from one part of the base to another. I had been looking forward to seeing Scapa Flow, this enormous base of the Home Fleet, where a German submarine had sneaked through all defenses and had sunk the *Ark Royal* at anchor. With a howling gale and heavy rain, there was little opportunity for sightseeing, but a kind officer explained in some detail how the German sub had managed to sink the ship. He did this without being able to disguise his admiration for the fine seamanship.

I visited a few other bases. Professionally, the trip was of but little interest except that I met my Norwegian colleague, Mr. Anda, and discussed with him meteorological problems related to the Stockholm airlift, missions to Norway, and air cover for convoys on their way from Iceland to Archangel. Anda was a first-class meteorologist, a theoretical communist, and a very practical patriot. As the war progressed, I gave Anda many awkward assignments.

After I had been less than two weeks at Dunstable a new situation developed. When the Nazis invaded Norway many Norwegian freighters were in Swedish waters; gradually a large number of them found their way to Gothenborg, where anti-Nazi feelings ran high. Early in the war the problem was to prevent these ships from falling into Nazi hands, but as the submarine warfare intensified it became imperative for Great Britain and Norway to secure these ships as replacements for convoy losses. Officially Sweden was a neutral country, but certain government officials in Stockholm and a few harbor officials in Gothenborg would, apparently, be willing to play innocent parts should the ships attempt to escape. However well the plot was organized, the escape was doomed to failure unless the ships could sail in heavy weather under a dense cover of low clouds lasting uninterruptedly for forty-eight hours. December, with less than six hours of daylight, was obviously to be preferred. With the blessings of Admiral Riiser-Larsen and Sir Nelson Johnson, I responded to a request from the Admiralty and worked with their forecasters to predict bad weather for the escape.

The Admiralty forecast center was housed in the Citadel, a new underground section of the Admiralty, supposed to be the most heavily protected shelter then in existence. Working here, it was easy to sense the pulse of the war. The weather was generally stormy, and on several occasions the cloud cover was favorable, but each time some hitch developed in Sweden, and the "man on the spot" decided against going. After about two weeks the whole operation had to be postponed.

In the meantime, other plans came to the fore. The convoy losses on the stretch between Iceland and Murmansk were increasing at an alarming rate. The ideal remedy seemed to be to cut off Norway at a narrow part near Vikna, about 100 miles below the Arctic Circle. However, as 1941 drew to a close, the overall military situation became rather fluid and Great Britain could ill afford to commit forces to a long-term scheme in northern Norway. As an experimental operation, on a much smaller scale, it was decided to attempt destruction of German ships in the Narvik area and to occupy a few of the Lofoten Islands to see how the Nazis would respond—fight or withdraw. In connection with this operation, a diversionary raid by a small Norwegian force was to be made on Vågsöy Island, some 500 miles farther to the south. In the initial phase, forecasts of the predominant weather for a period of four days were required, and the Admiralty invited me to assist.

The forecasts, as well as the operations, were very successful and much German shipping was destroyed. However, on account of unexpected developments in other theaters, the forces that had been landed were withdrawn before German intentions became fully clarified. Since my contributions to these forecasts were entirely advisory, credit belongs to the Admiralty forecasters. I was much impressed with their skill in handling maritime meteorological problems. Only two months before the outbreak of war in Europe, I had been elected president of the Maritime Commission of the IMO. I had recently developed close connections within the U.S. Navy, and I felt greatly encouraged by what I had seen in the Royal Navy. My ambition had long been to secure international support for an expansion of the networks of observing posts across the oceans, with the further aim of improving weather forecasting for oceanic as well as continental areas.

More than six weeks had passed since my arrival in England and I had had but little opportunity to attend to the upper-air problems at Dunstable. So I called on Sir Nelson and suggested that I would now need several months without interruption to formulate a program and to build up a group to develop analysis and forecasting techniques. The pressing problem was to find able scientists with a liking for practical problems—men who would think in terms of time rather than eternity. Sir Nelson did not need much urging. He had found two such men; they were both already at Dunstable; and more could be found as we expanded. And this was not all. Within a few weeks Professor Peter A. Sheppard, who was currently working in the bureaucratic headquarters atmosphere, would join me at Dunstable.

There are two versions of how this happened. My own recollection is that the initiative came from Sir Nelson. Peter (whose real name is Percival) insists that I invited him, that he requested the transfer, and that Sir Nelson was pleased to approve. The point of interest, however, is that our collaboration turned out to be extremely fruitful. Peter and I covered different but related fields of competence. When the war ended our group, which was then quite large, had developed a system of data processing and upper-air analysis that soon became standard throughout the world.

CHAPTER 9

The Upper-Air Group

There's nowt so queer as folk![49]
—Yorkshire proverb

No visitor to Dunstable could fail to notice some unique features of my upper-air group, as compared with the other and far more sedate sections of this agglomeration of meteorological activities. In January 1942, when I began to plan and organize the upper-air work, I found it wise to look around for young talent rather than trying to attract experienced forecasters. Weather forecasting is an art and a science. Although the art component derives from experience, not all experience is productive. Particularly in large organizations, weather forecasting easily becomes surrounded by a mass of routines, which tend toward undesirable standardization and often serve to stabilize stagnation. Since seniority rather than scientific knowledge and imagination tends to control promotions, one can hardly expect a personnel officer to make the best choices in trying to staff any new enterprise. The Meteorological Office's effort in upper-air forecasting had, until then, been next to negligible. The bulk of the forecasters had gained their experience from studies of sea level charts only, and I thought it preferable to search for potential, rather than realized, values.

However, in any category, there was an acute shortage of manpower. The director of personnel, who had the unenviable task of trying to meet all demands, seemed to be guided by the philosophy that any "odd person" who could not be advantageously posted to RAF stations was an especially gifted or budding scientist and ought to be posted to my group. Though hardly conducive to uniformity, this policy worked quite well in practice. Since there was a trickle of refugee meteorologists, I did not have to compete with other units for their services. Moreover, I was in a position to stimulate the influx of Norwegian meteorologists and technicians and to have their posting decided upon—to my group or to the field—as seemed most profitable. To maintain a progressive and balanced group and to ensure continuity at the end of the war, Sir Nelson saw to it that I got a fair number of young and gifted British scientists. As a result, we soon developed an exceptionally promising multinational group without any built-in inertia. The group was soon promoted to the status of a branch. At no time, before or after the war, have I had such devoted and stimulating collaborators.

At the high end of the scale was Professor Sheppard, whose job it was to monitor the Meteorological Office program in atmospheric physics, instrumental developments, sounding systems, and meteorological reconnaissance by aircraft. Peter was a model of clarity and order. He had the rare gift of knowing the precise meaning of words. Once in a while I have come across men who were such deep thinkers that they could not find words to express their thoughts; on other occasions I have heard men who were so eloquent that they could not find thoughts to go with their words. Peter had no such disability. I learned a great deal from my association with him.

At the other end of the scale was a refugee from an Allied country who had acquired the rank of sergeant, and, on the criteria developed by the director of personnel, had been posted to my group. As he reported for duty I went through the regular routine. When I inquired about his ambitions he stated, quite frankly, that he would like to become either a world-famous musician or a world-famous general. It occurred to me that if he possessed skill in writing music he would probably be good at plotting wind arrows; so plotting became his assignment; and he seemed quite happy. One of my technical officers had an artificial eye and was bothered by an infection, which caused him to wear an eye patch. My good sergeant found that an eye patch was useful as a symbol of manliness and acquired a patch for himself. His judgment in this matter was not too far off course for, with the patch, he looked like an anticipative reproduction of the famous Israeli general Moshe Dayan. This, perhaps, was as far as my sergeant ever reached in the field of generalship.

As time passed, my sergeant's plotting seemed to become more variable. At the same time, he developed a strange rhythmic practice of coordinated jumping and spitting. A report, through channels, to the personnel office seemed in order. A draft was prepared and locked in a drawer of my desk to which no one else had a key. Nevertheless, a rumor of the contents of my report leaked out before I had unlocked the drawer. Was there, after all, a spare key in existence? After some detective work it seemed plausible that a screwdriver, and not a key, had done the trick, and the screwdriver pointed to a certain Mr. X. If the desk were turned upside down, the top could be unscrewed to obtain access to the papers in the drawers. Since the screws looked quite worn, I found it advisable to change my security arrangements. Obviously, nothing useful would be gained by exposing the case. But in the next routine efficiency report I noted that Mr. X seemed to have a tendency to be unduly curious about matters not directly related to his work. In response the director of personnel explained in very friendly weasel words how my remark might harm the career of an employee of many years service. I was very pleased to alter my report to read: "Mr. X has far-reaching interests also outside his professional field of endeavor." The precise meaning of words?

On account of differences in background, customs, and outlooks, strange incidents occurred now and then. In any group of people working closely together a single incident may be irksome and leave behind much scar tissue, but if the incidents are sufficiently frequent and varied they may tend to make the group interesting and even attractive. Though one or two of the episodes could have been made to look ugly if dressed in the garb of standard procedures, bureaucracy had no place within the Upper-Air Branch. A number of minor incidents provided amusements, even outside the group. Most long-lived, perhaps, was the following. One morning, at the end of a night shift, one of us foreigners could not find his bicycle clips. All the girls, except a Miss X, searched the analysis room, turned over charts and diagrams, and created a lovely disorder—just to be helpful. Eventually, Miss X asked, "Have you felt in your pockets?" The response: "Here they are but, curse it twice, I have forgotten my bicycle!"

I have never looked to office bulletin boards for entertainment or other forms of uplift, but early one morning, as I entered the analysis room, I saw quite a crowd laughing at something pinned on the board. A notice had been posted, signed by one of the analysts, describing in drastic terms the competence of one of the data-handling supervisors, a well-balanced and charming lady whose work and sense of diplomacy I valued most highly. Without a word, I pulled the notice off the board and went into my private office. Everybody took the notice as a "foreign joke." The only thing that happened was that this graceful lady called on me, saying, "Please, do not do anything about that notice; he is such a nice person; and we are all very fond of him."

Something with an element of unpleasantness occurred one day, also on the night shift. Two analysts, one a foreigner (F) and one British (B), had worked through an irksome bombing raid. At the end of it, when the force was approaching home bases, something snapped. Both Mr. B—who had always been exceptionally enthusiastic about making all of us foreigners feel at home in England—and Mr. F were well liked by all. Nevertheless, things had gone wrong. As I returned to work in the morning I found on my desk a large weather chart on the back of which was written in enormous letters: "Sir, I have called Mr. B a swine, and I mean it! I am, Your obedient servant, F. . . ." That afternoon Mr. F and I found time for nine holes and a drink at the nearby Dunstable Golf Club. The upheaval soon took its place among other episodes in the arsenal of "foreign jokes." Best of all, Messrs. B and F remained the best of friends.

We were fortunate to have with us one of the best mixers I have ever come across; Hans Munkebye was his name. Hans was a Norwegian electronics expert and worked in the so-called Sferics Unit, *sferics* being a code name for atmospheric disturbances caused by electric discharges of different kinds. No country had come anywhere near the British in the development of sferics techniques. Together with three auxiliary stations, the unit provided bearings on sferics within a radius of 2,500 miles. My interest in these measurements derived from the information they provided on the vertical structure and movement of deep air masses. Sferics data from the North Atlantic were of particular interest, since customary weather reports from ships were extremely sparse. As soon as a sufficient volume of data had been obtained, Dr. F. A. Berson[50] and I made a comprehensive study of atmospherics in relation to fronts and air masses, which turned out to be a useful extension of Bjerknes's classical cyclone models. In our many uses of sferics observations Hans Munkebye was of very great help.

Hans was often in the limelight, and stories about him were not confined to the meteorological campus. One evening, as he was riding home from the golf club, with a lady sitting on his crossbar neatly supported between his arms, a policeman stopped him. Hans did not understand why he had been halted. When he inquired of the policeman, he was told, in a rebuking tone, "You are old enough to know!" Hans, who did not know that it was illegal in England for two persons to ride on a single bike, concluded that they had been halted for infringing the moral code. At this point the lady intervened and the officer requested no further information.

Hans was a brilliant bridge player. Some years before the war he and his Norwegian partner had beaten the famous bridge expert Ely Culbertson in an international competition. On another occasion Hans

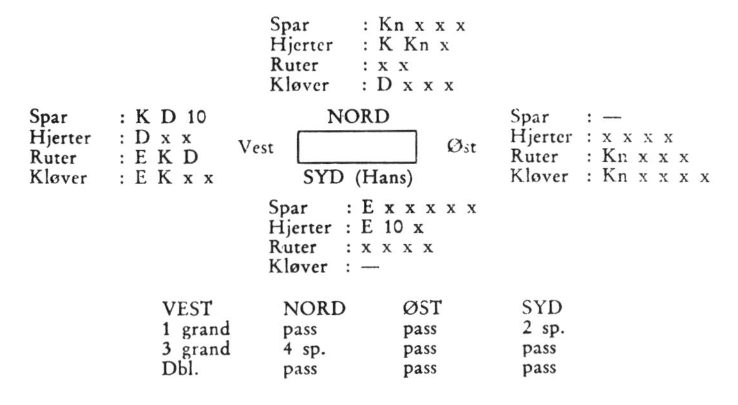

Fig. 9.1: Chart of bridge hands. Munkebye is South (*Syd*), Petterssen is North (*nord*).

invented a rare coup which, I believe, was referred to in one or two of the prewar texts on bridge. Some time in 1943, arrangements had been made for a so-called United Nations Bridge Competition in London, with the Greek ambassador as the host. Hans had found two Norwegian bridge players of class and, out of sheer kindness, he insisted that I join as his partner. I worked from early morning until teatime preparing upper-air analyses and forecasts for a bombing raid, and then handed it over to the next shift. Though one may hand over the work, the problems and worries tend to remain.

In the car, on the way up to London, Hans tried to impress upon me the rules and formalities of international bridge competitions. Though it all sounded very simple, things did not go too well. The Greeks were superior, the Dutch not far behind, and the Norwegians near the tail end. Then the hands shown in Fig. 9.1 were dealt, North–South (Norway) being vulnerable and the Dutch sitting East–West.

West crashed two diamond tricks and then led the king of clubs, which was trumped. By playing alternately hearts and diamonds from South and clubs from North, Hans had to surrender only one trump trick. When our colleague West inquired about our system of bidding, Hans passed the question to me. After some friendly discussion it transpired that I had not noticed West's opening bid (my mind had been on the upper winds for the raid) and thought that Hans had opened the fourth hand by forcing—top score for Norway, in both rooms. The real joke came the following weekend when a prominent Sunday paper reproduced the hands and commented, "North-South, being short of points, probably felt that action was called for, and North, taking a chance, landed his partner in four spades which was easily made." After one more useful blunder on my part, Norway came out with a respectable end score.[51]

The multinational character of our upper-air group varied from time to time. It reached a maximum late in the war when the U.S. Navy arranged for Professor George Haltiner and Commander Gene Bollay to work at Dunstable to become acquainted with our techniques. Mr. T. Henry of the Canadian Meteorological Service followed. As soon as Paris had been liberated, M. R. Pone of the French *Office National Météorologique* came for a prolonged visit and contributed generously to our research activities.

When the war in Europe ended, the group dispersed, I being the first to leave. A few of my colleagues remained in the Meteorological Office and rose to high position; others sought outlets for their talents in universities and comparable research organizations in Australia, Great Britain, Norway, Sweden, and the United States. Though three decades have intervened, personal contacts have been maintained—professionally as well as otherwise. Throughout these years I have considered myself fortunate to have had the opportunity to serve this group as their *magister lapidum*.[52]

Perhaps the most remarkable feature of the upper-air group was the subprofessional staff. Almost all of them were young ladies, all very beautiful, highly educated, well read, friendly, and hardworking. Most of the members of the professional staff were bachelors. I could not see how any of them could choose a bride in the upper-air group, for all potential brides were so uniformly attractive. However, as the saying goes: *überall wo es Menschen gibt, wird es gemenscht*.[53] By the time the war ended almost all were committed, and they lived happily ever after.

CHAPTER 10

Bombing and Research

. . . let me be hardened,
to suffice for this hard age.
—Nordahl Grieg

At the outset I hoped to build up a research and development group that could serve as a "think tank" in support of all forecasting activities then in existence. My plan was to analyze the atmosphere, layer by layer, up to great heights, combine these analyses with the data contained in the customary sea level charts, and try to determine how the various layers interact to produce the various weather systems. The philosophy underlying this approach had been developed by Vilhelm Bjerknes during the first decade of this century—long before upper-air sounding stations became established. In the thirties, Jack Bjerknes and Erik Palmén had used the method with great skill and elegance in their researches on upper-air structures. On a more limited scale, it had been used in the weather services of Norway and several countries in Central Europe, including Germany. Early in the war F. A. Berson brought the method to Dunstable, without much impact on their efforts in research and forecasting. As I saw it, our immediate task was to develop Bjerknes's method into routines and then progress from analysis of current situations to prediction of developments in the near future. However, external conditions imposed their blinkers, and, for many good reasons, I had to shrink my ambitions and become directly involved in operational forecasting.

Sir Nelson, always helpful and trusting, briefed me on the shape of things to come. He also invited me to sit with the Meteorological Research Council. We soon agreed that my group should concentrate its efforts on problems where useful results could be expected within a few months. Wind forecasting for night bombing and forecasting of night fogs at airfields in England topped our list of urgent items. The third item on the list was a kind of needle-in-the-haystack problem, namely, to discover the conditions associated with subsidence, or slow downward motions on a large scale in the free atmosphere. Typically, the speed of subsidence is about two inches a second, while the associated horizontal currents may be a thousand times stronger. In spite of their smallness, these sinking motions are important, for they account for the large and durable cloudless spaces aloft. There were a number of other problems that could be tackled as personnel became available. The list, in spite of its limitations, gave me something firm to hold onto. I felt we were off to a crisp start. As a result of this reorientation, my group became heavily involved with forecasting of upper winds for air operations. Bomber Command was by far our most demanding customer.

Up to the end of February 1943 British bombing of Germany had been sporadic and had been conducted on a very small scale. The guiding policy was to build airfields, produce equipment, train crews, and conserve aircraft until the force had become strong enough for a major effort. Though the bulk of the *Luftwaffe* was busy in the USSR and in the Mediterranean, the air defenses in the West were so strong that day bombing on any substantial scale was out of the question.

It was equally clear that night bombing, even of large targets, would be unprofitable until new navigational aids had been developed. A breakthrough in 1941 looked highly promising; all concerned were hopeful that an avalanche of night bombing could be released the following spring. By the end of February 1942, devices called "Gee boxes" had been installed in a substantial number of bombers. A Gee system consists of a master and two slave stations suitably arranged, transmitting radio pulses. The box measures the time lag between the pulses reaching the aircraft. From such lags, the crew computes the position of the craft. The useful range of the Gee system was about 350 miles.

Between March 3 and April 28, 1942, no less than eighteen major raids were launched. The first raid, which targeted the Renault factories on the outskirts of Paris, was highly successful. Beginner's luck? A large number of raids were directed at the heavy concentrations of industry in the Ruhr–Rhineland area, with Essen as the principal target. Cologne received considerable attention, as did several of the towns along the northwest coast of Germany. Rather spectacular raids were visited on Rostock and Lübeck, the latter being well beyond the Gee range.

An examination of the quality of our wind forecasts for these raids showed that none had been seriously in error; a few had been highly accurate. Though information on bombing performance was not routinely available at Dunstable, it was possible, in some cases, to compare the losses suffered by individual raids with our errors. The results were disappointing in that no relationship seemed to exist. Logically there ought to be a clear interrelation between the two factors.

In March and April 1942 a typical Gee-guided raid consisted of about 200 aircraft of a variety of types and performance characteristics, with some of the crews having but little experience in bombing. The squadrons took off from an assortment of airfields. Early in the operation they had to acquire a spatial arrangement, in the horizontal as well as along the vertical, that would enable them to pass over the target during a certain time interval and in the desired formation. The first wave of bombers would be much concerned with dropping flares and starting fires to create landmarks for the guidance of the following waves; the second wave would drop additional flares (to prolong the illumination) and loads of bombs; and the remaining force would then unload their cargo. The concentration of aircraft over the target was not excessive—about 120 per hour—and the duration of the raid was of the order of one hundred minutes. Since the planes were distributed over a considerable area and had to fly at different times, the desired formation could not be attained unless the winds were well predicted. Any significant error in the predicted winds would tend to disorganize the flight and the attack, with adverse effects on the bombing performance and the safety of the force.

Though the importance of the accuracy of our wind forecasts was quite clear, it seemed that the aberrations due to our errors were hard to detect among the deviations caused by many other factors. It was soon found that the Gee system, though an excellent aid in general navigation, was not sufficiently accurate to guide the force to the target unless visibility was sufficiently good for landmarks on the ground to be seen. For example, the raid on the Renault factories (March 3–4, 1942) was accomplished in excellent visibility and against rather weak defenses; as a result, the bombing was accurate, the damage heavy, and the losses small. Similar conditions prevailed at Lübeck (March 28–29) and at Rostock (April 23–24). In contrast, at Essen and other Ruhr centers, thick industrial haze was almost always present. The bombing was generally inaccurate, and the defenses, though quite strong, were handicapped by poor visibility. In this maze of disturbing influences, it was difficult to identify the effects of erroneous wind forecasts.

May 1942 was a relatively uneventful month with a much-reduced bombing tempo. Though some of the inactivity was due to unfavorable weather over Germany, there might well have been other reasons. With the nights becoming shorter, I concluded that no great activity could be expected until early autumn. I could not have been more wrong. During the May lull, Bomber Command prepared for raids on a vastly increased scale. The first of these raids was Operation Millennium. In the late evening of May 30, a force of about 1,050 bombers set out from fifty-odd airfields, with the ancient city of Cologne as its target. The moon was full, the sky almost cloudless, and Cologne received a devastating blow. Compared with earlier raids the average concentration of bombers over the target had increased fivefold. The raid lasted ninety-eight minutes. The first aircraft over the target arrived seven minutes early; and the last left exactly on time. Clearly such precision could not have been attained unless the wind forecasts had been highly accurate.[54]

Operation Millennium was followed by a few somewhat smaller raids, with Essen as the principal target. As on earlier occasions, industrial haze intervened and the bombing was not highly effective. Something better than the Gee system was needed to pinpoint attacks on targets covered by clouds of thick haze.

Several improvements in bombing techniques and technical aids to navigation were introduced in 1942 to support bombing raids during overcast conditions. The first innovation was the establishment, within Bomber Command, of a Pathfinder Force under the vigorous command of Air Vice Marshal D. C. T. Bennett, a master navigator who, in October 1940, had captained the first flight of bomber aircraft from Newfoundland to England. Now, in 1942, highly skilled officers and men were selected and organized into a single corps d'elite to find, illuminate, and mark the targets to be bombed. The Pathfinders would drop flares and incendiaries to provide a pattern that could be recognized by the bombing force. Close on their heels would follow the first wave of bombers, the fire-raisers, with their loads of incendiaries. After them would come the main force with their cargo of incendiaries and explosives.

Under this scheme, since the various components of the raid would fly at different heights, the winds in the whole air space below about 35,000 feet would be of interest. Normally, and particularly in winter, this meant that the Pathfinders, the fire-raisers, and the main bomber force would be flying in wind regimes of different speed and direction, the variations with height depending almost exclusively on the temperature distribution below flight level. Coordination of the movement of the different components of the raid became a matter of utmost importance. Satisfactory results could be obtained only if we could deliver upper-air forecasts of higher accuracy than we had done in the past.

The importance of strict coordination is readily understood. For example, if the Pathfinders were to arrive early, the enemy air defense would be alerted, flares and markers might be extinguished, decoy fires started, fighters readied to meet the bombing force, and losses would be high. On the other hand, if the Pathfinders arrived late, the bombing force would have nothing specific to bomb and would be an easy prey to the defense. Strict coordination was necessary to safeguard the compactness of the force and to ensure a sudden arrival, a quick delivery, and a snappy departure. Compactness was needed also on the outward and homeward journeys to maneuver between heavily defended areas. The importance of accurate upper wind forecasts now stood out with abundant clarity.

Soon after the Pathfinder Force began to operate, two major technical innovations were introduced to remedy the shortcomings of the Gee system. The "Oboe" system consists of two ground stations of which one (the "Mouse") directs a radio pulse over the center of the target. As long as the aircraft stays on the beam the pilot will hear an oboe-like note. If he strays to port or starboard, he will hear different signals. At the same time, the pulse is returned to a second ground station (the "Cat") which, at intervals, computes the distance along the path of the aircraft and, at the end, identifies the target. The range of the Oboe system was no greater than that of the Gee, but the former could take the aircraft to within a few hundred yards of the target, whereas the latter might be as much as 6,000 or more yards off. Unfortunately, Oboe service could be provided only to individual aircraft. For this reason, the effort was concentrated in the Pathfinder Force.

The second, and equally important, technical innovation was the H2S (Home Sweet Home), which permitted vertical scanning. Essentially, the H2S is a crude self-contained television transmitter and receiver installed in the aircraft. Since prominent terrain features such as rivers, bridges, railway yards, and the like reflect the energy of radar waves with contrasting intensities, a crude picture of the terrain under the aircraft will appear on the screen. While the Oboe equipment had to be reserved for the Pathfinders, the H2S found wider application. A number of aircraft in selected bomber groups were equipped with H2S for wind-finding purposes. An aircraft flying on a straight and level course would convert two successive vertical H2S fixes into ground speed. By subtracting the ground speed from the recorded air speed, the average wind experienced by the aircraft could be obtained.

Though the method was simple in principle, the accuracy of individual observations was generally low. Not often could a bomber maintain a straight and level course and provide strictly vertical beams over the terrain features of interest. Other sources of errors also contributed. As a result, there was normally a large scatter among the winds observed by neighboring aircraft. This scatter had to be eliminated to find a representative wind.

The introduction of the Oboe and H2S systems sharpened rather than relaxed the demands for accuracy in our wind forecasts. In a typical raid the squadrons would take off from a variety of airfields, find their way to the rendezvous, and assume their allotted place in the compact structure of the bombing force. Highly accurate wind forecasts were much in demand for this phase of the operation. From the rendezvous to the coast of the continent no H2S fixes were possible. Again, accurate forecasts were needed to preserve the compactness of the force. From the coastline to the target, H2S winds could be obtained in considerable numbers. To remove the scatter, the winds were radioed back to Dunstable and plotted on a central diagram. Obviously wild cases were eliminated. From the remainder, a "center of gravity" was determined and compared with the flight forecast and observations that had become available since the forecast was issued. Next, Dunstable would radio back, via some control station, a drift correction for the last ten-minute interval and two forecast corrections, each for one-minute intervals. This hectic work continued, six times each hour, until the force crossed the coastline on its homeward journey. The objective was to get the compact bomber force, with the Pathfinders and various appendages, to the target within three minutes of the desired time, and to maneuver between heavily defended areas on the outward as well as on the return journeys.

Sometimes things went wrong. On the whole, however, the Oboe service, a conspectus of electronic devices for jamming and other purposes, the H2S winds, our general forecasts, and our ultra-short-range drift corrections contributed to keeping the losses down to a level above which the growth of the bombing force would become endangered. After Oboe and H2S had become routine, it became clear that, in general, the losses increased in proportion to the errors in our wind forecasts.

In spite of a heavy and steadily expanding load of routine duties, it was possible to accommodate some research work. We had set out to develop a relatively new field. Since so little had been done before, it was not difficult to discover problems. The upper-air group was imbued with curiosity, yet the pressing demands of Bomber Command and other customers caused us to keep our feet squarely on the ground. To maintain liaison with reality was no problem; indeed, to work under pressure provided many valuable stimuli. Other conditions were not unfavorable. Since Dunstable was an exceptionally dull place, work on research problems served well to fill in spare time. It was indeed pleasing to see young scientists drift back to our office barracks to work during off-duty hours. It soon became a routine for me to return to work from 8:00 to 11:00 in the evening.

Since our prognostic analyses were distributed to the field, the group prepared a number of technical memoranda to keep the recipients informed of our techniques. These reports were distributed within the British and American services. The always alert U.S. Navy Meteorological Service saw to it that they reached field stations as far out as American Forces went. Though these reports and memoranda did not discuss fundamentals, they served a useful purpose by describing new procedures and preparing the way for international adoption of standards soon after the war ended.

We soon found that there was an urgent need for an accurate and mutually consistent set of physical constants and functions used in meteorology. Only if all nations were to adopt such a set would the data from different parts of the world form a satisfactory basis for analysis and prediction. Professor Sheppard took charge of this work. With the assistance of N. Herlofson, one of my Norwegian assistants, this rather onerous task was brought to an early finish. In 1947 the IMO adopted the set, reproduced the report, and accorded it wide distribution to weather services, universities, and other research organizations.

Our analyses of the air currents in the upper troposphere (25,000–30,000 feet above the ground) revealed the existence of a highly concentrated and very strong westerly current, meandering from the Rocky Mountains to the Soviet Union, and presumably around the whole world. The existence of such a current had been suggested by the Bergen meteorologists as early as 1933,[55] but it had not been charted until 1942. At Dunstable, Berson was keenly interested and marveled at its concentration and strength, but since the Meteorological Office chose to classify all weather reports, no description of this remarkable phenomenon could be attempted.[56]

Fog research, conducted by W. C. Swinbank, became productive beyond expectations and resulted in a set of prediction diagrams that proved of great value in guiding returning bombers to bases in the early morning hours when fog is most frequent in England. This research was also productive in a wider sense. It inspired C. H. B. Priestley and Swinbank to reexamine and extend the classical theory of turbulent motion, thus providing a more solid basis for studies of the all-important processes associated with the redistribution of heat, moisture, and momentum in the atmosphere.[57] Soon after the end of the war, Priestley and Swinbank went to Australia, where they were later joined by Berson. Since then Australia, with only ten million people, has become accepted as one of the "great powers" in scientific meteorology.

My own highly varied work at Dunstable was frequently interrupted by assignments to special jobs. The demand for technical memoranda to field stations was always pressing. Once in a while I found time to ready a research paper for publication. The vast majority of my research did not reach beyond the "rough-note" stage. In early May 1945, I left Dunstable with a large case full of such notes, all centered on upper-air structures and their relations to forecasting. These notes, then, became the vade mecum that I could nibble at during the following seven years when management of meteorological operations became my primary duty. It was not until 1952, when I returned to academic work, that I could organize, expand, and use my notes [see chapter 25]. The result was a two-volume text on *Weather Analysis and Forecasting* (1952) that soon found wide distribution, with translations into Japanese and Russian.

Chapter 11

A Stab at the *Tirpitz*

When the forts of folly fall.
—Matthew Arnold, *The Last Word*

Although I was loaned to the British Meteorology Office and took instruction from them, Admiral Riiser-Larsen, as commander in chief of the Norwegian Air Forces, was, in a technical sense, my boss. I therefore made a point of keeping him informed, in general terms, about my activities. Gradually a kind of friendship developed between us, based on mutual respect and a sense of sharing in broad outlooks, rather than on social activities and in-service gossip and rumors.

Riiser, as he liked to be called by friends, had made his early career in the air branch of the Norwegian Navy and advanced to the rank of rear admiral. Later, as a result of a reorganization of the armed forces, he acquired the rank of major general. Internationally, his fame rested on his feats as a pioneer in Arctic and Antarctic exploration. His knowledge of conventional aircraft, dirigibles, icebreakers, and similar craft was both wide and deep. His calmness in critical situations and the general aura of his personality instilled confidence and respect in all but a few pedants who felt that his leadership owed too much to inspiration and not nearly enough to staff work and time-honored procedures. In some cases the pedants had a valid point, but they overlooked Riiser's vision and drive.

Sometime in February 1942, I called on Riiser. I wanted to enlist his influence in obtaining seats on the Stockholm airlift for two Norwegian meteorologists; they were wasting time as refugees in Sweden while I was short of talent for my upper-air work. It so happened that King Haakon was visiting the Norwegian headquarters when I arrived. Riiser lost no time in having me invited into his presence. Much to my surprise, the king seemed well informed about my activities. He reminded me of Article Eight in Hitler's ultimatum, demanding that the Norwegian weather service be placed at the disposal of the *Luftwaffe*, and he seemed genuinely pleased that Norwegian meteorologists were so readily received in the British service.

I had had the honor of meeting the king on several occasions before the war. Now, as he usually did, he began to joke about the weather and my "professor science." The winter, which was nearing its end, had been raw and sunless. And when Riiser-Larsen remarked on this and compared it with Oslo in November (before the crisp snow begins to brighten the landscape), the King laughed away this little bit of gloom by observing, "Then this is the longest autumn I have ever experienced."

Riiser, though he liked to command and make decisions, really had the heart and mind of a poet. His optimism was often boundless. On one occasion I heard him saying that Hitler really lost the war in Norway, and the rest would not take very long. Now, he seemed to be rather pessimistic and genuinely concerned about recent events. Although the structural balance of the German navy was irretrievably lost during the invasion of Norway, the submarine situation was developing from bad to worse. I thought things were gradually improving, but Riiser went on to describe the German advances toward the oil fields in Southeast Asia and pointed out that the shortage of oil might not be a limiting factor much longer. Although the Germans might be stopped short of the oil fields, the Japanese were spreading southward with enormous speed, perhaps focusing the American war effort there rather than against Hitler and his hordes. British and Norwegian ships, laden with munitions, were queuing in British ports, unable to proceed to Archangel for lack of escorts. Clearly, the first thing to do was to eliminate the German battleship *Tirpitz*; thereafter many things might change.

Riiser's customary optimism was obviously returning. I was glad, for I had a train to catch to reach Dunstable in time to get ready for the night shift. Reviewing the conversation in the train, I felt I had been a poor listener. Riiser had spoken as if he wished to tell me something; as if he had a message but did not find a suitable occasion to deliver it. Perhaps my mind had strayed too often to the problems that I would have to face later in the year when the large concentrated bombing operations on Germany were expected to begin.

Some days later I received an invitation from Air Marshal Baldwin, then chief of Bomber Command, to call on him at his headquarters at High Wycombe. Expecting to hear about bigger and better raids on industrial centers in Germany, I reviewed my problems and was prepared to explain what my group could and could not do. I also rehearsed the arrangements that would be necessary to improve our ability to forecast the winds in the upper atmosphere.

A staff car took me to High Wycombe, where I was cordially received. Coffee was offered, as if my Scandinavian habits were well known. Then the air marshal showed me some reconnaissance photographs of a winter landscape: a large fjord with frozen bays, surrounded by wooded mountains and some scattered farms. "Do you know this place?" was his question. Expecting no catch, and being eager to be helpful, I said, after some scrutiny, "This is Lofjorden, about twenty-five miles east of Trondheim." From the smile on the air marshal's face I realized that I had had my leg pulled. Obviously, he knew all about the photographs. His question (like the coffee) was only meant as an introductory pleasantry. My knowledge of the area was, however, far beyond what could reasonably have been expected. In the first place, I had lived in Trondheim throughout my teens and once or twice camped not far from Lofjorden. Furthermore, as a sergeant in the Norwegian army, I had served with a unit that kept guard over the German auxiliary cruiser *Berlin*, which had violated Norwegian neutrality and had been interned in Lofjorden toward the end of the First World War. Being responsible for provisioning the unit, I had traveled frequently to neighboring villages and towns. I knew the place in great detail.

The air marshal handed me a close-up of Lofjorden, which is a bay of Åsenfjord, which, in turn, is a branch of the very much larger Trondheim Fjord. He asked me if I could see anything unusual. After some time, and not without hesitation, I had to confess that I could not recall a tiny island close to a small peninsula. A magnifying glass revealed the *Tirpitz*, anchored close to the steep but rather smooth side of a rocky promontory. She was heavily camouflaged with pine and fir trees and a variety of gear.

The general plan of a forthcoming operation was now explained to me and details were added later by the commander of the attacking force. The monster was too strong for the bombs then in existence, but buoyant mines or depth charges dropped on the rocky side of the promontory close to the ship might find their way to the soft underbelly below the torpedo-protecting belt. Two earlier attempts at the *Tirpitz* had failed because fog, real or artificial, was found to cover the ship and the surrounding area. To drop "new things" through a fog would achieve nothing but giving away information. The forthcoming attack had to be handled with utmost care.

To ensure surprise the assault had to be made by night. There were to be two waves of bombers; a squadron of high-flying Lancasters was to go in first, at 15,000 feet or more, to knock out the anti-aircraft guns and create confusion. Then, as soon as operationally feasible, a squadron of Halifaxes was to follow up with a "suicide attack" at very low level, a few hundred feet or so, dropping depth charges and mines.

The chances of success were not thought to be high (though the expected losses were), but the *Tirpitz* was a target of great importance. The job given to the Halifaxes was one of extreme precision bombing. To ensure accurate aiming, the sky had to be completely clear, with a full moon so high that no shadow would fall on the ship, the near side of the promontory, or the narrow strip of water between them. Only three nights in the lunar cycle would satisfy the nonmeteorological requirements. But worst of all, at this time of year and so close to the Arctic Circle, the nights are shortening at a rapid rate. Only one full moon, the one at the end of April, would be available. If this opportunity were lost, it would mean waiting until the autumn and explaining the delay to the prime minister, who was already impatient.

The operation was to be carried out by Four Group led by Air Marshal Carr, with headquarters in York. The two squadrons involved were stationed at Lossiemouth, on the north coast of Scotland. For security reasons the meteorological service could not be handled, as on earlier occasions, through customary channels. Would I be willing to go with Air Marshal Carr and his staff to Lossiemouth to "press the button" for the operation? I was more than willing. The briefing had left no doubt about the importance of the operation and the fruits that would follow success.

There is a peculiar fascination in the forecasting of natural phenomena. I am not referring to the prediction of sunrise and sunset, the seasons, and that sort of thing: mere repetitions, like the tick of a clock. I am referring to the forecasting of variables, of phenomena that at one time do not even exist; phenomena that flare up, grow to a peak, then fade away and, eventually, become buried in a graveyard called meteorological archives. The true forecaster will revisit the graveyard, dig out old corpses, add little frills here and there to the original analyses, and try to squeeze a little more information out of unwilling victims. And so the storehouse of experience is enriched.

Forecasting is exciting. The competitive instinct, or whatever it may be called, swells and drives the forecaster into action—action with caution. A storm is neither a friend nor a foe. It cannot be controlled, praised, or punished. It must not be allowed to enter unannounced. The forecaster's will must prevail. The challenge and the elation are largely lost in general-purpose forecasting—often dreary routine; predictions without an address, predictions without a known purpose, perhaps with no purpose at all. But the scientific challenge and the elation that stem from an exalted sense of purpose and usefulness are enormously enriched when you are called on to forecast for a specific and important operation, one whose purpose and structure have been fully explained to you. You have then become a member of a team: You exchange knowledge (and sometimes wisdom), you ask and answer questions, and you feel that the team as a whole must come out victorious.

The visit to Four Group was exceedingly pleasant. The chief meteorological officer and his staff went out of their way to make me feel welcome. Here, there were no toes waiting to be stepped upon; the *spiritus loci*[58] was very much to my liking.

On arrival at Lossiemouth I had several surprises. In addition to the air marshal and his chief of staff there was a youngish British admiral and several high-ranking officers. Also present was Sir Archibald Sinclair, who, during the First World War, had been deputy commander of Churchill's battalion in France and was now secretary of the air in Churchill's government. Rumor had it that the prime minister was on the telephone at least once a day, pressing for action. Clearly, the *Tirpitz* was a high-priority target.

Almost three months had passed since the *Tirpitz*, the giant battleship that never saw much of the high seas, found shelter in the Trondheim area, and no punishment had yet been inflicted. I did not know then that Churchill had started to press for an attack three days after the *Tirpitz* anchored at Lofjorden. On January 25, 1942, the prime minister wrote to the Chief of Staffs Committee:

> The presence of *Tirpitz* at Trondheim has now been known for three days. The destruction or even the crippling of this ship is the greatest event at sea at the present time. No other target is comparable to it. . . . The entire naval situation throughout the world would be altered, and the naval command in the Pacific would be regained.[59]

Clearly the pressures on Sir Archibald, the air marshal, and others must have been enormous. Sir Archibald had probably more music to face than the rest. There were also a few disappointments. The Germans had recently changed their ciphers and our very capable Station

X had not yet broken the keys. As a result, few reports from enemy-held territory were available, and none at all from the Trondheim area.

In some respects the meteorological situation seemed manageable. We had regular balloon soundings from Shetland, which, among other things, showed that a high pressure area (or anticyclone) was situated north of Scotland. The temperature distribution was such that it reached up to great heights. Excellent news: The general situation was stable and could be expected to persist. Equally important, the presence of a deep anticyclone suggested persistent subsidence (slow downward motion, normally about ten feet per minute) with relatively dry and cloudless air aloft. However, by the same token, it was also likely that low clouds would persist, with fog forming where the air from the snow-covered land met with the warmer ocean air. The low-level conditions would be determined not by the high-level circulation but by the winds in the very lowest layer. Since we had no reports from the boundary layer, I was not prepared to engage in guesswork.

This initial survey led to two clear conclusions. First, the upper-air situation could hardly be more favorable and predictable. Second, without additional observations, nothing could be said about clouds and visibility at low levels. The Lancasters and the Halifaxes might be affected by two different weather regimes; one known and the other unknown. To recommend for or against attack was clearly out of the question.

I called on the air marshal and found him in conversation with Sir Archibald and the admiral. They seemed most anxious to hear my advice, but were somewhat surprised when I said that I needed reconnaissance aircraft to provide low-level observations from the Trondheim area. Though my synoptic information would be useful, photographic reconnaissance was much to be preferred. Here was I, a civilian with an outlandish accent and sadly lacking in Englishmanship, talking to men of exalted rank. However, my request, which was stated in what nowadays might be called "a low profile," was received with understanding. After a short period of silence, the air marshal explained in a friendly tone that such planes simply were not available. Though I was not unprepared for a negative, my reply was quite spontaneous. I just said, "Then I don't see how I can help you." Sir Archibald, tall, erect, and looking very impressive in his enormous wing collar, suggested that other bomber groups might be able to help.

After a brief absence, the air marshal returned, looking quite pleased: Reconnaissance had been arranged. But now I found myself in a very embarrassing situation. I had not thought of making it clear that the pilots had to land at Lossiemouth so that I could brief them, explain my worries, and tell them what to look for. To send out planes "to look for weather" would serve no useful purpose. I was truly sorry to have to raise this problem after arrangements had been made, but I felt I had no choice. The air marshal put me at ease. New arrangements were made, the pilots were briefed, and returned the following morning with reports and excellent pictures.

A wide fog bank, with a sharp edge toward the northwest, covered much of the Trondheim Fjord and the land southeastward. Both Lofjorden and Trondheim were covered; the fog was deep, for none of the hills, which I knew so well, could be seen. Above the fog the sky was perfectly clear: The upper-air situation had behaved. The sea was dead calm with no indication of wind. Without wind there was no reason to assume that the fog would disperse. The first of the three possible nights had to be ruled out.

Reconnaissance was arranged for the early hours next morning, and little change was found. The sky was as clear as before, and though the edge of the fog had receded somewhat, there was no indication of even light winds. The second night, too, had to be turned down.

Reactions were mixed. Although nothing useful had been achieved, abortive missions had been avoided and, as far as we knew, the secret had been well kept. When I considered the minute cost of the reconnaissance as compared with the rest of the effort and the gigantic benefits that might derive from a successful attack, I was glad that I had asked for the tools I needed and grateful for what I had received. But what about the third and only night left?

The last occasion with moonlight was the night of April 27–28. Reconnaissance in the early morning of April 27 showed that the fog bank had receded farther, but Trondheim and Lofjorden were still covered; the fog was deep and the sea calm. However, something new had happened. A ship had anchored just off the fog edge at Trondheim, probably 50–100 yards from the harbor installations. Smoke from its funnel showed a slight drift in an offshore direction. I took this to indicate that a light drainage wind had begun to make its influence felt. In midwinter this could lead to the formation of arctic sea smoke, a fog that develops when bitter winter air moves over open water. Such fogs are not infrequent in the Trondheim Fjord; however, in April the temperature contrasts are moderate, and sea smoke is quite unlikely. Moreover, the fog bank had receded landward, indicating that no such fog could be expected. Most likely drainage winds would dissolve the existing fog.

Although the tilt of the smoke was rather flimsy evidence, the process itself was well understood;[60] the questions were ones of degree and duration. The reconnaissance showed that the upper-air situation had not changed, and there was nothing to indicate that it would be disturbed within the near future. Adding up the pros and cons, reviewing for the umpteenth time certain consequences, and, on and off, thinking of Wing Commander Bennett who would lead the Halifax attack, I decided to make a positive recommendation—without qualification.

It has been said that a meteorologist should limit his advice to a purely meteorological statement. This is obviously sound if the structure of the operation is not fully known to him for, as in other fields, a little knowledge has but little advantage over complete ignorance. However, if the forecaster has been let in on the operational plans in sufficient detail, as I had been on this occasion, discussion and interpretation are far more useful than a cold bureaucratic message; circumstances, rather than circulars, must decide.

For security reasons, and certainly not as an indication of inhospitality, my "office" was a tiny shack far out in the field. It was not easy to spread out my many maps to study the time sequence of events. The air marshal asked me to brief him and his companions on the meteorological situation. I had wall space for only four maps on a narrow vertical strip by the door. Unfortunately, I had hung the last and most important map at the lower end of the strip and the four of us sufficed to crowd the room. I went quickly over the history of the case and ended with the latest chart, the one that almost touched the floor. Sir Archibald, who showed interest in details and had many searching questions to ask, assumed a squatting position, sitting on his heels, and I followed suit. I was impressed with the cogency of his questions and his extraordinary ability to obtain meaningful information. I had planned a rather short summary briefing, with details to be explained to the commanders of the task forces. Instead, I had been led into a lengthy question-and-answer session in a not-too-comfortable position. I was then forty-four years old and in fine trim, so when the briefing came to an end I got up without any difficulty. Sir Archibald did not do too badly. The somewhat bulky air marshal and the admiral remained standing throughout our session. It was evident that in substance my briefing had gone well. The operation was on.

Two of the three possible nights had been lost, but the reasons for the inactivity were completely documented through the reconnaissance photographs and other reports. So far, nothing had gone wrong. Hopes for the third and last night were obviously high. However, there were formidable unknowns: the strength of the ship's armor, the efficiency of the mines, the accuracy of the drops, and the behavior of the fog. None of us went so far as to say "poor *Tirpitz*."

From an operational briefing later in the day, I understood that about thirty minutes had been allowed between the high-level attack by the Lancasters and the close-up assault by Bennett's squadron. Phrasing my words with care, so as not to reflect on nonmeteorological matters, I interpolated that my confidence in the forecast of the upper winds was extremely high northward to beyond Shetland and high for the remainder. After all, the high pressure area had stayed put for some time and had been "well tried out." I went on to suggest that one might wish to shorten the interval between the two attacks. If the interval was too long, the Germans might manage to cover the area with artificial fog, since four or five "fog factories" were known to exist around the bay. My point was appreciated, but opinion was against shortening the interval. Operations of this type, over such large distances, cannot normally be timed with minute accuracy. Nothing but failure would follow should the Lancasters arrive too late. Although on this particular occasion a shortening of the interval would have served well, the statistics of air warfare spoke against it. The timing of coordinated operations at different levels depends upon the vertical shear wind, the forecasting of which remained a nasty problem; it lasted longer than the war.

Although the green light was on, no unusual activity could be seen. Apparently everything had been planned with care and everybody knew his job. Since routine meteorological work did not now take much of my time, I used the opportunity to see the base. As I walked about it occurred to me that many of the young men I had met might be killed during the following night. This was my first direct contact with fighting men and my thoughts began to stray over the whole field of the horrors of war. What was it Lee had said when he rode away from Gettysburg? "We might get too fond of war if it were less gruesome," or something to that effect. Why did Erasmus leave so little impression on

European culture while Luther, a loosely educated fanatic, caught the imagination of the masses? And did not gentle Fröding write:

> Though you condemn war and bloodshed,
> Knowing the sufferings so well,
> You, nevertheless, stand in awe
> Of the glory of battle.

Futile speculations, reminiscent of Ibsen's onion: You peel off layer after layer and in the end you find nothing. I returned to my quarters.

My bedroom was next to that of Wing Commander Bennett and I dropped in to see him. My knowledge of the Lofjorden area might be useful to him. He had spent the night before with his family; now he was getting his equipment ready for the coming night. It seemed just plain routine. When I noticed his long underwear spread out on his bed, I thought that pure wool would have been warmer and far more absorbent. As a young man, I had lived in snow huts or caves for a total of about four weeks. I had driven with reindeer and pulkas through winter storms; I had led a skiing party through the Norwegian, Finnish, and Swedish parts of Lapland; and I had learned to fear clamminess more than cold. A snow cave can be quite comfortable unless you are wet or sweaty when you settle in. Should any of the men be shot down at Lofjorden they would have to try to find their way through snow and wilderness to reach the Swedish border, a distance of forty-odd miles. Without skis they might have to spend more than one night in the open. However, I saw no point in raising such questions at this time. I was familiar with Bennett's pioneering work in the early days of transoceanic flights, but this, too, seemed out of place. So I told him that late reports did not indicate anything unfavorable and that my confidence in the forecast remained high. Unless something unusual should turn up, I would come to his briefing and might then have fresh reports from Shetland.

Something did turn up, but I still heard the tail end of his talk: a magnificent summary of objectives, methods, possible snags, and, above all, what to do if shot down. I had come to think of Bennett as a quiet, studious person, a civilian expert in uniform. He now appeared as a real leader of men—leadership through analysis, clarity, and skill, with no trace of showmanship. Some of his men might not have returned had they not been so well instructed.

During the evening meal I obtained the impression that the air marshal had received a message from the prime minister to be read to the men before takeoff. The planes and crews were now scattered over a wide area and there was no time to assemble them for a reading. The air marshal very kindly invited me to join him and the admiral in a brief visit to Bennett's squadron, which had the unenviable task of flying in low over the ship. Thoughts of loss of life came back to me, but in a somewhat different context. While the air marshal spoke to individual crews, I kept looking for faces that reflected forethought. Not a single face! They were all cheerful—or pretended to be so. The more distant men who could not be heard made unmistakable signs to indicate that they were going to have a really good go at the enemy. I returned enormously impressed with the relaxed discipline and the real courage of the men known by the acronym RAF.

After takeoff there was little to do but wait and hope. The air marshal and the admiral invited me to a game of snooker, but since such games had formed no part of my upbringing, I could only offer to observe and admire. It transpired that Air Marshal Carr was a close friend of Sir Hubert Wilkins. As young men they had joined efforts in an expedition to the Antarctic. Elderly aunts and other ladies of similar wealth and vintage had financed the expedition, wholly or partly. The objective of the expedition, he told them, was to explore the "m e t e o r o l o g i c a l" and "c l i m a t o l o g i c a l" conditions of the Antarctic. Carr had acquired a way of pronouncing these words in a manner that must have sounded highly scientific and convincing to elderly ladies who trusted in youth and had money to spare. I was fairly familiar with the history of Arctic and Antarctic exploration but had not come across references to the findings of Carr or Wilkins. Since there can be no really good answer to a stupid question, I refrained from asking. I had not seen Sir Hubert since his visit to Tromsø after his transpolar flight in April 1928. I was very pleased to hear that he was still active and doing useful work at both ends of the earth's axis. I was saddened to hear that his handsome companion, Carl Ben Eielson, had been killed in Alaska.

On the morning of April 28 the squadrons returned. The weather had been ideal: pure polar air, brilliant moon, no fog, and no shadow on the target. The Lancasters had come as a complete surprise and dropped their cargoes. The Halifaxes arrived on time, but by then beams of fog were seen to stream from land-based installations toward the ship, and very soon, the bay filled. The approximate position of the *Tirpitz* was marked by black smoke penetrating the white screen of fog, but the accuracy of the aiming must have been much reduced. Not much damage was done to the ship. Flak was very heavy, suggesting that the bombing by the Lancasters had not been highly effective. Most of the mines and depth charges dropped by the Halifaxes did not find their way to the soft underbelly; those that did seemed to have been too weak to do much damage. Reconnaissance after the event seemed to indicate a slight list toward port, that is, toward the promontory. The *Tirpitz* seemed to have had a very uncomfortable night but had not suffered too badly.

Five aircraft did not return. The losses were certainly lighter than expected. One of the missing planes was Bennett's. Apparently he had pressed his attack very hard and had been hit repeatedly by flak; when the engine caught fire the crew bailed out. The woods around the bay were heavily guarded, but some men escaped in the confusion. Bennett and his radio operator were the first to reach the Swedish border. Here they presented themselves as escaped prisoners of war, which, according to convention, entitled them to repatriation rather than internment. Bennett was soon back with the RAF. He was promoted to air vice marshal and placed in command of the Pathfinder Force of Bomber Command.

I met Bennett twice after the *Tirpitz* operation, once at Bomber Command in early 1943 when we tried to sharpen the procedures for wind finding and the use of wind forecasts for bombing operations, and again in 1948, at the headquarters of the International Civil Aviation Organization in Montreal. He had retired from the RAF and was back in civil aviation, where his career had started. From later reading I learned that he at one time ventured into politics, was elected as a Liberal to the House of Commons, and had crossed swords with Churchill. Apparently, this master at air navigation found maneuvering between political reefs rather difficult. He will long be remembered for his excellent work in navigation and as chief of the Pathfinder Force. One does not readily forget a man like Bennett.

But we must return to Lossiemouth. While Four Group was bombing the *Tirpitz*, the *Luftwaffe* had a nasty stab at York. Though much damage had been inflicted, Air Marshal Carr's family was safe and, as far as we knew, so was his headquarters staff. As work came to an end and I was about to look up train schedules, the air marshal put his broad hand on my shoulder and said, "I have a seat for you in my plane." Fortunately, I had only a small bag and a briefcase, for the plane could barely hold three men, and there was some gear to be accommodated. My next surprise was when I was told where to sit. I had a glorious view over the landscape, the east coast of Scotland, bleak in places but beautiful everywhere.

At York the air marshal said goodbye and rushed off to see his family and the results of the bombing. A plane took the admiral and myself to London where a car was waiting at the airport. The admiral insisted on taking me to my hotel. There we parted, expressing hopes that we might meet again. About ten weeks later, as I was standing outside the old War Department building in Washington, trying to catch a taxi, the admiral approached, with the same objective. We chatted for a while. He was then on his way to the Far East where recent losses had been heavy. The *Tirpitz* was still afloat, tying up the Home Fleet.

Although the meteorological part of the operation had turned out to be a model affair it seems right to say that success was not so much due to skill in three-dimensional forecasting as to the circumstances that I had obtained meteorological reconnaissance to provide me with indispensable data. From comments by Air Marshal Carr and others, I realized that the weather reconnaissance had served a second important purpose. The reasons for the decision not to attack on the first two nights of full moon and the circumstances attending the attack on the third night could be fully documented. Useful lessons could be learned and a closer point of departure for further actions had been established. The operation as a whole had been extremely well managed; the air marshal had every reason to be pleased. The meager results were due to the equipment and the techniques then available compared with the enormous strength of the *Tirpitz*.

In reviewing what I had seen and heard, it seemed that night bombing, though effective against large areas, cannot be relied upon against small targets. In spite of her bulk, the *Tirpitz* was a very small target. Perhaps the prime minister had a point when (as I learned much later) he wrote in his aforementioned minutes to the Joint Chiefs of Staffs Committee:

> No doubt it is better to wait for moonlight for a night attack, but moonlight attacks are not comparable with day attacks.

And also:

> There must be no lack of cooperation between Bomber Command, the Fleet Air Arm, and aircraft carriers. A plan should be made to attack both with carrier-borne torpedo aircraft and with heavy bombers by daylight or at dawn.

The *Tirpitz* lingered on in Lofjorden for a few weeks and then moved to Kåfjord, a branch of the Altenfjord, on the west coast of Lapland. She was now 700 miles farther removed from the RAF and, by the same token, that much closer to our convoys to Russia. In September 1943 she was badly damaged, but not crippled, by midget submarines and men with courage "beyond the call of duty." But the *Tirpitz* remained afloat and a potential menace until the autumn of 1944. She was then completely crippled by repeated attacks, at dawn and in daylight, through complex collaboration between Bomber Command, the Fleet Air Arm, aircraft carriers, and torpedo aircraft, as suggested early in 1942 by the prime minister. In the meantime, the bombs had grown from a few hundred pounds to 10,000 pounds. The very end of the giant came on November 11, 1944, near Tromsø, where she was beached and used as a fort. I last saw the *Tirpitz* in April 1946. There she was, with a gaping hole in her exposed underbelly, a true monument to human folly.

CHAPTER 12

Mr. Pyke and the Mastery of the Snow

C'est magnifique, mais ce n'est pas la guerre.
[It is splendid, but it is not war.]
—Maréchal Bosquet[61]

On a sunny day in 1937, a tall, slim Englishman with a saintly beard walked along the tortuous path from Eidfjord to the Hardanger-Jökelen [glacier] and saw, for the first time in his life, the great white stillness, the vast expanses of snow and ice, the Hardanger Vidda [plateau] in all its glory. He stopped at the crest, looked back down the dark and forbidding valley, and then again turned to the fields of glittering snow. Being as richly endowed with imagination as he was with patriotism, the Englishman murmured to himself: Britannia rules the waves, we invented the locomotive and the tank, we are now about to lead the work in the air, it is time for us to master the fourth element—the snow. The author of this monologue, whispered in the great white stillness of the lofty mountains of Norway, was Mr. Geoffrey Nathaniel Pyke.

I shall long remember Mr. Pyke. Indeed, few people who met him will forget him, though they may remember him for different reasons. For my part, Mr. Pyke and his obsession with snow involved me in a curious diversion from my work at Dunstable.

Pyke was brilliant, but utterly unorthodox in all he did. His originality had become apparent early in his career. When the First World War broke out, Pyke tried to join the army but was deemed physically unfit. He then acquired a fake American passport and an assignment as a reporter operating out of Berlin, sending regular reports to the London newspaper *Daily Chronicle*. Eventually he was caught and interned in a camp north of Munich. After a while he confided in a fellow English prisoner and suggested that they return home. His compatriot was willing, provided that Pyke could make "safe arrangements." Pyke did precisely that: He acquired a text on statistics, made 8,000 observations on the behavior of the guards, and analyzed the data as the text prescribed. Next, he made an appointment with a friend in Rotterdam. Then, in broad daylight, he and his companion escaped, found their way to neutral Holland, and kept their appointment.

Back in London, Pyke was asked by his newspaper to write an article describing his escape. Instead he wrote on the theory of escaping. According to Pyke, to escape successfully, one needs complete knowledge of the geography of the camp and the behavior pattern of the guards. In addition, two important things are required: a hat and a cow. The hat is needed to make you feel respectable, to forget that you were a lowly prisoner; it gives you self-confidence. The cow is needed to make you appear an unhurried local person wherever you may be. To escape is as simple as that. Twenty-five years later, while Pyke and I were in Washington, a tabloid carried a small news item: A French prisoner had escaped from a German camp and arrived in Nantes leading a cow. Whether or not he wore a hat was not mentioned, but it seemed probable that he had read or heard of Pyke's theory of escaping. Pyke could justly call himself a professional thinker.

Pyke had an inventive mind. In 1938, when Hitler's intentions seemed abundantly clear, Pyke began to ponder war-prevention schemes. In the end he decided to try to achieve his objective by exposing Hitler as a fool. He would prove to the world that the majority of the German people were against Hitler and did not want to go to war. His plan was to conduct a secret opinion poll in the Third Reich and then publish the result. With some helpers in Great Britain and Germany, Pyke worked out all the details, organized all impedimenta and, with his little task force and much luggage, went to Berlin. After careful study and discussions with his German helpers, Pyke decided that, unfortunately, it was too late to start a nationwide poll. Since he and his English helpers had to return, they had to find a way to take much of their paraphernalia with them. Hitler and Goebbels were known to believe that England would stay out of an armed conflict on the Continent. So Pyke invented a rescue stratagem for himself and his group. He managed to bluff Goebbels into using his services to further a friendship-with-England scheme that helped Pyke get out of Germany with all that ought not be left behind. This ingenuity in handling Nazi officials was astounding, especially as Pyke was himself a Jew. When he told this story he would laugh and say that he was the ideal, the prototypical Jew, and that Nazi Jew-baiters could recognize only common Jews.

When war broke out, Mr. Pyke devoted his energies to a series of war-winning schemes. Since his first glimpse of the Hardanger Vidda, he had accumulated much data on snow, and even consulted one of my books, *Weather Analysis and Forecasting*, which, in a chapter on air-mass transformations, contains a map showing the duration of effective snow cover in the Northern Hemisphere. By stretching definitions somewhat, and without making allowance for customary exploitation factors, Pyke had come to the somewhat exaggerated conclusion that whoever mastered the snow could control no less than seventy percent of Europe.

Pyke regarded snow as a raw material that could be processed in various ways. He soon succeeded in producing a hard bullet-repelling substance called Pykerite by freezing a mixture that, essentially, consisted of water and sawdust. He also toyed with the idea of using hollow icebergs to transport war supplies from America to Great Britain. There were other war-winning schemes, all logical in general structure and appealing particularly to minds that consider science as something mysterious, without distinguishing clearly between a mystery and a miracle. But the immediate and burning problem for Pyke was the logistic mastery of the snow. The first step toward this goal was to design and manufacture a snow vehicle with near-miraculous performance characteristics.

Pyke found his way to Lord Louis Mountbatten, who, in 1941, had succeeded Admiral Lord Keyes as chief of combined operations—the famous British Commandos. Mountbatten had begun to surround himself with a brain trust, including some men of exceptional gifts. Lord Louis listened to Pyke's argument: The war was going to be fought and won on many fronts; a bullet fired in Lapland would cost Hitler as much as fifty bullets fired in Central Europe; through the mastery of the snow Hitler could be forced to disperse his strength. To achieve this, a snow vehicle with extraordinary capabilities had to be developed. Time was short.

Pyke, like most amateurs, was a believer in destruction as a way of winning the war though, of course, sophisticated methods of destruction were much to be preferred. The high-head hydroelectric power plants that he had seen in Norway and in the Italian Alps ought to be eliminated from Hitler's war production machinery. Pyke conceived the "snake," an ingenious contraption that would do precisely that. Pyke had little background in technology, but given his background in statistics and data manipulation, he was more than just a dreamer. His fundamental thesis was that humans can think their way through any problem. He liked to describe in some detail the chain of mental processes that led him to his different war-winning schemes. In the case of the snake, the vital part of a hydroelectric power station is the turbine. The turbine can best be destroyed by explosives, and the best place for the explosive substance is inside the turbine. How does one get this substance into the turbine? One lets the water itself carry it to the destination. Hence, the snake must be able to carry explosives, swim into the tubes, be carried by the water, and explode on contact with the turbine. The rest is just engineering!

In late March 1942 General George C. Marshall and Mr. Harry Hopkins, representing President Roosevelt, visited London, primarily to press for sympathy and support for the Bolero Project, the invasion of the continent of Europe in the spring of 1943. During their many conferences with British colleagues, the question of diversionary operations came up more than once. Through such operations Hitler could be forced to spread his forces more thinly, giving Bolero a better chance of success. In line with this sort of thinking Lord Louis informed General Marshall that Combined Operations had been studying a scheme that might well become an ideal diversion. The idea was to land a glider-borne force with snow vehicles in the Norwegian mountains to destroy power stations, creating confusion and suspicion of bigger things to come, thus attracting German forces from the Continent. If this were to assist Bolero, a snow vehicle, far superior to anything then in existence, would have to be designed and produced within about ten months. This could be done only in the United States. General Marshall readily agreed to explore the possibilities.

Churchill was known to be opposed to an early cross-Channel invasion and was much in favor of peripheral actions at any time. On April 9, 1942, exactly two years after the Nazi assault on Norway, Pyke went to Chequers to brief the prime minister, General Marshall, Mr. Harry Hopkins, and others on his grand scheme, which soon acquired the project name Plough. In preparation for the briefing at Chequers, Pyke had prepared a memorandum under the heading "Mastery of Snow," with an outline of strategic objectives:

1. After thorough consideration we have come to the conclusion that in the coming winter the United Nations must be masters of the snows. Our studies of the strategic situation have led us to the concept of snow as a fourth element, a sea which flows over most of Europe each year and which usually tends to act as a brake on military operations. We must obtain mastery of snow as we have of the sea. By mastery of the snow we mean the possession of cross-country snow machines which would make almost all parts of territories such as Norway and the Carpathians accessible by such vehicles. This would

enable us to move over snow at speeds greater than that of the enemy and to go where he cannot follow. We include particularly the ability to go up the gradients prevalent in areas of operations and also the ability to go through forests.

2. The development of a machine capable of performing these functions will bring nearer than any other means at present at our disposal the prospect of:
   (a) Eliminating Norway as an economic asset to Germany by the destruction of the hydroelectric stations. Forty-nine percent of the hydroelectricity in Norway is concentrated in fourteen stations.
   (b) Simultaneously with (a) above, destroying a large portion of the oil refining capacity and a considerable proportion of the oil producing capacity of Rumania. With adequate preparation, this might be done at a single blow.
   (c) Compelling the enemy to lock up large forces in Norway in an anti-sabotage role.
   (d) Preparing the way for an eventual re-occupation of Norway, thereby forming a link with Russia.
   (e) The destruction of German bases in northern Norway, which threaten our supply line to Russia.
   (f) On the southern slopes of the Alps, the destruction of as much as seventy percent of the Italian hydroelectric power.

General Marshall, on his return to Washington, summarized the discussions at Chequers, emphasized the high-level enthusiasm in London, and set up a preliminary search and feasibility study. It was now late April; summer was not far off. If Bolero were to be helped, a near-miraculous snow vehicle had to be designed, tested, and manufactured in large numbers. Equipment of various kinds had to be provided, men had to be trained in a highly unorthodox kind of warfare, military plans had to be made, and, not to be forgotten, aircraft had to be made available to move the force. All this had to be accomplished within eight or, at most, nine months.

Meetings of American and British representatives were held, bottlenecks identified, and shortcuts explored. The Office of Scientific Research and Development (OSRD), headed by Roosevelt's science adviser, Dr. Vannevar Bush, was willing to assume responsibility for the development work, but there were certain conditions attached:

1. The manufacturer to give top priority to work on the vehicle;
2. The army to assist in collecting specimens of all existing snow vehicles;

3. The Departments of State and War to lend fullest support;
4. Dr. Sverre Petterssen to be released from the British Meteorological Service to work on the project; and
5. Dr. Herman Mark, of Brooklyn Polytechnic Institute, to be retained as a consultant.

So this was how I became involved in work on the mastery of the snow.

One day in late May Sir Nelson telephoned. Lord Louis Mountbatten wanted to see me; it would be advisable for me to be prepared for travel. I had not the faintest idea what it was all about. Sometimes it is unwise to ask questions. My guess was a novel type of commando raid on Norway, something much bigger than the Vågsöy and Lofoten raids in late 1941, with which I had been loosely associated. Perhaps I would have to go north to a base in Scotland. It might be wise, I thought, to have warm clothing. On arrival in London I checked in with Sir Nelson. He had no information beyond that I might be away for two or three weeks. Knowing that there was some difference between English and Norwegian styles, I asked Sir Nelson how one should address Lord Louis, a son of Prince Louis of Battenburg, and a close relative of his Britannic Majesty and other ruling monarchs. The answer was simple: "Sir" is enough on most occasions. In the taxi down to Combined Operations Headquarters I went through a brief but adequate "Sir-ing" exercise.

At the appointed stroke of the clock I was taken to Lord Louis' office. No pomp, obviously a busy place; a handsome and, apparently, brilliant man, fast-moving, no waster of time. After a number of pleasantries he came to the point: "We are interested in snow and ice; our colleagues in Washington have asked for your help on an important joint project. I am going to Washington in two or three days time, and will tell you more about the problem when we meet there. Miss So-and-So, in the anteroom, will make arrangements for you."

I had many good reasons for wishing to visit America, but far stronger arguments for staying on my job at Dunstable. We had recently done quite well at a series of major bombing raids on easy German targets. In all these raids the winds aloft had been moderate or light, but winter storms might raise new problems. My group was in good shape, but we needed experience. On the other hand, the nights were getting short and no large-scale bombing could be expected until September. An absence of two or three weeks would make little difference. The nearest I came to a negative was to say that I knew next to nothing about crystallography of snow and ice and, therefore, someone else

Fig. 12.1: Petterssen in uniform as a lieutenant colonel in the Royal Norwegian Air Force.

might be more useful. The answer, with a smile, was much to the point: "You are needed in Washington for what our colleagues there know that you know." I liked the phrase "our colleagues." I was reminded of Eliza Doolittle in *Pygmalion*.

I arrived in Washington on June 3, 1942, wearing the hastily acquired uniform of a lieutenant colonel in the Royal Norwegian Air Force (Fig. 12.1), the purpose of the uniform being to facilitate travel and to mollify those who think that only typists and men in uniform can keep secrets. Lord Louis had just arrived and Pyke had been there several weeks. Though Lord Louis was fully aware of the many problems that had to be solved and that time was not on our side, his vocabulary did not include the word "difficult." His drive was irresistible and his enthusiasm readily rubbed off on all who worked with him. I was told, in passing, that the president and the prime minister expected to be kept informed of progress. I took this to mean that red tape and money matters would not stand in our way.

The immediate problem was to find Pyke. He did not seem to fit into any of the organization charts that adorned the walls of government offices in Washington. Pyke lived and worked in his hotel room and often refused to be disturbed. However, my arrival was expected and the reception was very cordial. Pyke, who seemed pleased to tell me that he was responsible for bringing me to Washington, was in no hurry to tell me about the nature of my work. He was writing some notes on psychological warfare. He informed me that there was much I had to learn before we could discuss his thoughts on how to master the snow.

A brief survey showed that about a dozen different kinds of snow vehicles existed, but none of them was capable of doing what the Plough project demanded. It seemed to me that the requirements were even stricter than those indicated in the memorandum prepared for the briefing at Chequers. The machine had to be able to tackle bare ground, cross railway lines and similar obstacles, and, if possible, swim short distances in water. All existing vehicles had been built, more or less as had the first aircraft, on intuition. Most of them were fairly well suited for some specific job. However, if we were to outsmart all of these early contraptions and meet the many requirements of the Plough project, we had to know much more about the mechanical properties of snow.

Snow is a very elusive substance. Its properties at any time depend on its history from the time it formed in the cloud, through the many changes suffered on the ground brought about by evaporation, changing temperature, wind, and sunshine. Moreover, since the snow on the ground tomorrow may be quite different from that of today, a military commander must know something about tomorrow. On further analysis the problem of the mastery of the snow was found to be far more complex than anticipated. Given snow and time, I felt that we might come up with a solution, but the summer heat was already oppressive and the whole job had to be completed by Christmas. I felt very much like Nansen's Eskimo who had come down to the Greenland strandflats in March to cut grass and found that he had to sit down on a rock and wait while the snow melted and the grass grew. Like the Eskimo, we were about six months out of phase with the seasons; unlike the Eskimo, we were chased by men in high places, men whose job it was to ignore difficulties and get things done. To describe the obvious, least of all to Lord Louis, would serve no useful purpose. However, I did insist that we needed preparatory work in three stages: one, to study snow properties in order to derive a useful classification of major types; two, to determine the significant mechanical properties of each type and to translate these into vehicle design; and three, to determine associations between changes of snow properties and varying weather conditions.

Lord Louis needed little urging. Since there was snow (of sorts) in the high mountains of Alaska, I acquired a generous supply of blank travel orders, White House priority, and set out for Juneau, changing planes in Seattle. The Pan Am ticket clerk, a very charming young lady, was so impressed with my priority rating and my Norwegian uniform that she arranged for an invitation to a Norwegian dinner party to save me from the boredom of waiting for departure. While I was at the party Pan Am reached me with a message from Lord Louis. On account of the Japanese attack on the Aleutian Islands, the commanding general in Alaska could not support the intended experiment; instead, I was to go to a certain hotel in San Francisco where one of his representatives, Col. E. A. M. Wedderburn, would put me in the picture. My priority soon provided me with a seat on the plane, which arrived in San Francisco early next morning. Great was my surprise when I found that the person in charge of the engineering work was none other than Palmer Cosslett Putnam, the engineer who had conceived and directed the wind turbine project [see chapter 5]. It was pleasant, also, to meet Lord Louis' trusted assistant, Col. Wedderburn, a Scot in tartan. Wedderburn was a highly cultured and friendly person whose difficult job it was to please Pyke as well as our American colleagues. Pyke was not there; he rarely mixed with ordinary people.

Putnam lost no time in briefing me on the status of the enterprise. As usual, he emphasized the fascinations rather than the obstacles. We discussed at great length the urgent need for a hasty but meaningful research effort, and we compiled a short list of experts in soil mechanics. In 1942, as far as snow research was concerned, scientists were not much ahead of the Eskimos, whereas soil mechanics was quite well developed. With luck, we might find a shortcut to useful knowledge.

The day wore on. I had behind me three busy days and two nights with hardly any sleep and I was beginning to look forward to the proverbial shortcut to health, wealth, and wisdom. Then came a second message from Lord Louis: He was leaving from Ottawa next day and would like to see me at Montreal's Dorval Airport. My travel orders and White House priority sufficed to provide me with a seat on the evening plane to New York. Putnam felt very sorry for me and gave me two of his—as I discovered later—very potent sleeping pills, suggesting that one might not be enough. Though I had no experience with sleeping pills, I was, optimistically, looking forward to a long undisturbed rest. On the plane I happened to be seated next to a female, an incessant talker, completely hint-repelling, heavily scented and painted. Once upon a time, before she had acquired her present bulk and aggressiveness, she might have been attractive and acceptable, but time, which is supposed to heal all wounds, will also wound all heels. Anyway, in my judgment (which is not always perfect) she was nothing but a bothersome superannuated succubus, trying out her tricks on a man who was not yet asleep. The best I could do was talk her into swallowing one of my pills and, after a safe interval, I took the other one myself.

One pill was enough. I woke up after landing in Chicago. The captain and the stewardess were trying to shake me back to life. Their hope was to transfer me to an ambulance and send me to a hospital; they could not be responsible for a passenger who did not respond to landings and takeoffs, shouts, shakings, etc. After I showed my priority papers, explaining the circumstances, and mentioned my appointment with Sir Lord Somebody, they agreed to let me continue. I slept on and off, changed planes in New York, and reached Dorval on time, still feeling unrefreshed, drowsy, and much in need of a full morning routine.

The meeting was a flop. There really was not much to talk about, and I was certainly no source of inspiration. Also, the heat was oppressive. While Lord Louis, in his admiral's uniform, was a model of neatness in dress and person, Pyke had taken considerable liberty with his attire. His long English shirttails were flapping in the wind, and his endeavors to tuck them in were not altogether successful. The farewell ceremony was brief. Pyke and I had a few technical points to discuss; then I shook him off, found a hotel, and slept for eighteen hours on end.

Back in Washington Putnam and I conferred with experts in soil mechanics and found that they had but little to offer. Their terminology was very much the same as ours but the physical substances—soil and snow—were so different that we came no closer to information that could take us to engineering applications. As a by-product I became more firmly convinced that we had to find snow.

While in Washington we met Dr. Herman Mark, a leading figure in the science and technology of long molecules, fibers, and plastics. Like myself, Mark was a recent immigrant to the United States, but our backgrounds were quite different. He came from a distinguished Austrian family and had been educated in Vienna and Berlin. On his entrance to university his father had given him a platinum spoon for use in certain chemical experiments. He had served in the Austrian Army during the Kaiser War and had done some snow research for Austrian ski troops in the eastern Alps. In Berlin he had lived through the postwar depression and had had many interesting experiences, including the repeated pawning and retrieval of his precious spoon. Mark was an easygoing, friendly, warm person; outwardly a lederhosen type of Austrian with all signs of comfortable *Schlamperei* [sloppiness], but a model of method in his scientific work. Mark was a real find—the first bit of luck that I had been able to identify. But snow remained a problem.

Lord Louis had briefed the prime minister on the progress of the Plough project: the vehicle development, the snow problem, and the need to establish a specialized military force. During his visit to Ottawa, the Canadian authorities had agreed to join in the project. More detailed arrangements with the Canadians were made during a visit to Ottawa by Col. Robert J. Frederick (representing General Marshall), Mr. Pyke, and myself. The National Research Council was keenly interested in the development of snow vehicles for civilian as well as military purposes, and I chalked up their cooperation as a second piece of good luck. But we needed snow—not in season, but *now*!

There was snow in the Andes and in New Zealand, and no one would forgive me if any opportunity were missed. Since time was at a premium, a type of tracked vehicle, to be called the Weasel, had been tentatively decided on. The Studebaker Corporation was already making experimental models. The snow research we had in mind included testing new models as well as existing vehicles. Therefore, we needed not only snow, but snow where we could do our research and testing in such a manner that no message would reach Berlin. In 1926, Professor Joel Hildebrand, Dean of the Graduate School at the University of California, Berkeley, had led the American ski team to the Olympic games at St. Moritz. What would seem more reasonable than for the two of us to take a skiing vacation in the high Andes? Several places near Santiago seemed particularly inviting. We might see something of interest there. If not, New Zealand was always available.

On our way south we stopped for the night at Cali, Colombia. The passport office let Hildebrand through without any questioning. But I, traveling on a Norwegian passport, and carrying, in case of need, credentials as a consultant for a New York-based industrial concern, was questioned in detail about the purpose of my journey, the places I intended to visit, any friends or relatives in South America, and so on. After some time, when I began to wonder what next, the officer lowered his voice and, very agreeably, offered me compensation for spying a little—"jost a leetle"—here and there and reporting to him, personally, on my return. He emphasized the personal element and I concluded that he was not curious on behalf of his own government. There was no reason why I should not oblige, so I said I would keep my eyes open. He very kindly let me have his name and address and we parted the best of friends. While in Santiago I made up an interesting story, but changed circumstances caused me to return via Buenos Aires and Bahia.

Joel and I found Santiago very interesting and exceedingly pleasant. Our contacts were most cooperative and hospitable. There was snow in the nearby mountains, masses of it, crusted by sun and wind. There was also the promise of occasional soft snow in many accessible places where work could be carried out in great comfort. Joel and I spent a very pleasant weekend with a fabulously wealthy industrialist at his "little place" about 8,000 feet up in the Andes. This little place was equipped with a private ski tow, servants, workshops, two Picassos, a grand piano, and a private broadcasting station. The owner had lived in London for several years and was fully conversant with the technique of understatement. The weekend before our visit Walt Disney had been a guest. It was not easy for Joel and myself to invent something suitable for the guest book—something to go immediately after an autographed drawing of Mickey Mouse.

However, Chile was not a suitable place for the work we had in mind. A large component of the population was of German extraction. Although the majority was believed to be pro-Western, the remainder could not be ignored. So Joel returned to Berkeley and I went on to Buenos Aires.

In the early thirties the director of the Argentine Meteorological Service, Dr. Galmarini, had asked if I could spare one of my meteorologists to head the forecasting division. I had recommended a certain Mr. Alf Maurstad, who was duly appointed. Alf was a brilliant young physicist, a bachelor whose philosophy of life included two articles of faith:

One, unnecessary work is unnecessary; two, for a wife and a radio set, it is wise to wait for next year's model. I decided to stop off in Buenos Aires and, heavily camouflaging the purpose, find out if the eastern Andes held out any promise. The results were clearly negative.

Buenos Aires, however, is a place one cannot leave without formalities. Dr. Galmarini had developed a program, the Norwegian ambassador had a party already arranged, and in the years of Argentine wealth, no one could come to Buenos Aires without eating a steak at the Cabaña. When all this had been done, the local office of Pan Am was unable to find a seat for me on any of their planes to Washington. I concluded that I had become a suspect person, and that the local office had to yield to pressures that came to them, via high places, from the German Embassy. After three days of idleness I called at the American Embassy where I was lucky enough to meet the agricultural attaché, who happened to be of Norwegian ancestry. I told him the truth, nothing but the truth, but not all the truth, and asked him to relay a message to General Moses of the General Staff in Washington. Next day it so happened that a seat was available for me—in a half-empty plane—and we flew on to Miami, stopping in many places on the way.

I soon discovered that I had a companion who, apparently, spoke German and Spanish fluently and English moderately well. We stopped for the night at Belém, Brazil. After supper I went for a stroll. My companion followed a few horse lengths behind me. At one time when I thought I had shaken him off, I slipped into a cinema. The place was hot and the show boring. In the first interval I got up to leave and saw my companion sitting three rows behind my seat. He was not a hard-boiled professional. When I looked him straight in the face and said, "Good evening," a faint blush was his only response. As far as I could see the chase ended in Miami.

Germany was not the only great power concerned with my innocent love of snow and skiing. Four months later, the British Foreign Office requested the Norwegian Ministry of Foreign Affairs to provide them with an account of my doings in South America. Not wishing to trust honest Norwegian officials with boring details, I told the ministry that I had been on a skiing vacation in the Andes, and that Lord Louis Mountbatten and General R. G. Moses, General Staff G-4, Washington, D.C., were familiar with the details of my vacation. That seemed to close the British Foreign Office file.

Flying back to Washington, I had plenty of time to think about my failure to find snow in South America. While New Zealand had many inviting features, I was convinced that the research work could not be separated from the testing, and that logistic support could not be arranged without much loss of time if the testing were to be done in New Zealand. I then decided to fly to Toronto to see my good friend Dr. Andrew Thomson, director of the Canadian Meteorological Service. Very cunningly he arranged for a quick staff study of the skiing on the glacier in British Columbia. Old slushy snow was of course present all through the summer; fresh snow could be expected in the last week of August. We could do much useful work on these types of snow before wind crusts and sastrugi[62] began to form, probably in mid-September. Old hard crusts were not much of a problem. We now had a really convenient place for research and testing, within easy reach of the Canadian Pacific Railway.

With incredible speed the U.S. Army established a camp of quonset huts near the dome of the glacier, collected specimens of existing types of vehicles, and provided the necessary services. In the meantime, Dr. Mark had perfected all instruments needed for measurements of snow properties, and the Studebaker Corporation had made several tentative models. Obviously the Studebaker Corporation acted on business principles, but equally obvious were the patriotism and enthusiasm of their engineers and staff. Now we had a laboratory, an almost unlimited testing field, a company of soldiers, and all the facilities and services we wanted. And best of all, Mark had made arrangements to spend much of his time on the glacier, thus allowing me to spend about two-thirds of my time with the planners in Washington.

If the Plough project were to be operated in Norway, someone had to provide detailed information on the geography of snow cover and snow properties and their normal variations during the snowy season. I was fortunate enough to obtain the assistance of two compatriots: Dr. Harald U. Sverdrup, world-famous arctic explorer and geophysicist, professor of oceanography, and director of Scripps Institution of Oceanography; and Dr. Gunnar Randers, a young astrophysicist, then at Yerkes Observatory. To avoid waste of time in obtaining security clearances, I rented a pleasant house in Washington. There, Sverdrup and Randers, with the assistance of Mrs. Randers, completed a detailed study of snow and ice in Norway within three weeks.

Speed was essential, and speed had been achieved, in the War Department, on the glacier, in Studebaker's factories, and in my house on

Rodman Street. Vehicle models had been made, tested, and discarded, and new models produced. In mid-October a number of high-ranking officers gathered on the glacier. Dr. Mark had measured the snow properties. Using the diagrams developed for the purpose, we predicted that our vehicle, named the Weasel, would climb slopes up to twenty-six degrees. The Weasel did better than predicted, and the runner-up managed slopes of eighteen degrees.

The first version of the Weasel was limited in size to be suitable for air drops; it was amphibious and managed very well in relatively calm water. Later versions, fairly large landlubbers, are still the best all-purpose snow vehicles in existence; they have performed well in the Arctic as well as in the Antarctic. They can also tackle mud, sand, and hard roads. I felt relieved that my detour into work on snow and ice had come to an end.

What happened to the Plough project? The briefing at Chequers on April 9 had overlooked one or two details. Did the Norwegian government in exile wish to have their power stations destroyed? And if they did, what would the underground at home say when the war was over? Questions of this kind were slow in taking shape. When answers began to emerge the contours were diffuse. It was like looking at something strange through a bad fault in a windowpane. The problem of aircraft support was never really tackled, and interservice rivalries seemed to have weakened the forces behind the project. In the end, all difficulties brought forth a decision that had the elegant garb of wisdom: The objectives could be achieved with far less destruction. Plough was replaced by one or two sabotage schemes.

Although the original Plough project fizzled out, the Weasel, and its amphibious version, the Duck, did excellent work in many different parts of the world. The First Special Service Force, created to operate the Plough project and commanded by General Frederick, accumulated an extraordinary record of military feats and, unavoidably, casualties. It was this force, equipped with Ducks, that surprised the Japanese and reconquered Kiska and Amchitka Islands in the Aleutians, thus ending the enemy occupation of American soil. The same force, with Weasels, landed on the beaches of Morocco and fought at Oran, landed near Naples and fought in the mud and winter rains in Italy, landed and fought at Anzio, went on to fight in Corsica and on the Îles d'Hyères, and joined in the invasion of the French Riviera. Its last parade and casing of colors took place in early December 1944. Since the war had now grown into routing operations of armies and army groups, this small force of 2,400 daredevils had become redundant.

What happened to Pyke? Early in the planning of the Plough project a decision had to be made as to the general type of vehicle to be developed. An overriding consideration was the method of propulsion. There were three choices: the air-screw, the Archimedean screw, and the track. The first of these had to be ruled out since we were interested only in a heavy-duty machine. The Archimedean screw was an interesting vehicle. Essentially, it consisted of two large horizontal cylinders with steeply rising screw treads and a motor that turned the cylinders. Its compaction properties were next to ideal, regardless of the type of snow, and it could be made to pull a sled with a heavy load. Its speed and climbing ability were limited, but its main shortcoming was its inability to cross bare ground, railway tracks, and similar obstacles. Nor could it be made into an amphibious vehicle. On the other hand, a tracked type of vehicle could operate on bare ground as well as on snow, it could easily be propelled on water, and, properly powered, it could be made to climb slopes of considerable steepness. Its weakness was related to compaction: The track has a tendency to churn up certain types of snow. It was hoped that this drawback could be minimized by improvements in the design of the track.

Various combinations were also considered. Especially tempting was the Canadian Bombardier, a bus-type vehicle with skis in front and track under the rear part. This is an excellent vehicle on hard roads, but in the softer varieties of snow the skis cause the front end to float while the track tends to dig in under the rear; in soft snow it tends to dig its own grave. Another possibility was something resembling the snow toboggan, with track under the middle. The gliding surface gives even flotation but, under certain conditions, the track churns up the snow and propulsion is lost.

Unfortunately, Pyke had taken a firm liking to the Archimedean screw and did not readily accept engineering opinion. The complexities of my status now became a problem: I was a Norwegian employee loaned to the British air ministry, which, in turn, had loaned me to Lord Louis. On this line of reasoning I could be regarded as a person working for Pyke. On the other hand, my American connections considered me a member of their team, or at least as a person to whom they could have direct access. Although I had no mandate from any Norwegian authority, I did not feel that I could go along with any plan that, in my judgment, was not in the best interests of Norway. In any case I felt it right to side with the Americans in the preference for a tracked vehicle. This, together with Pyke's unconventional methods of transacting business, widened the rift between him and his American colleagues and eventually led to his separation from the Plough project and his concentration on other war-winning schemes.

Pyke's separation from the Plough project became a turning point in his strange career. He continued with several other schemes, including one in anti-submarine warfare, but none came to fruition. After the Allies' landing in Normandy, men in high places lost interest in new schemes. For Pyke, who once ascended to influence at the summit, the trend was now definitely downhill. Illness added to his problems. In 1948 this remarkable genius did the only orthodox thing he ever did in his life: He swallowed vulgar poison.

CHAPTER 13

The Failure of Operation Freshman

Mistakes are always initial.
—Pavese[63]

While I was in Washington in July 1942, my friend and benefactor at MIT, Dean Moorland, looked me up and invited me to dinner in a place where security would be no problem.[64] Speaking for an unidentified group of physicists and engineers, he told me as much as he thought necessary about the American effort in atomic energy technology. There was good evidence that the Nazis were also active in this field, and no one could say who would win the race. Most disturbing was a rumor that the Germans were planning an enormous expansion of the Norske Hydro complex of hydroelectricity at Rjukan, about fifty miles west of Oslo. There was another story, rather more vague, about a similar expansion at certain falls near Lake Enare in northern Finland. If there were any truth in these rumors it could only mean that the Germans were preparing to produce large amounts of heavy water which, in turn, could be taken to indicate that their experimental work on the development of an atom bomb was nearing its end. What Moorland wanted to know was how we could find out, quickly and safely, whether these rumors represented plausible extrapolations. Had construction work actually begun, there would have been no difficulty in finding evidence, but at the present stage the problem seemed very much one of mind-reading.

Rather sooner than expected a message through an underground connection clarified the situation. At Rjukan blueprints did exist, but there was nothing to indicate serious intentions. With the Nazi love of paperwork the blueprints and plans were likely to be nothing but file-fill or, more probably, just a personal career development project of some ambitious Nazi officer. Nothing definite could be found out about Enare, but considering climatic conditions, logistics, proximity to the USSR, and other factors, it seemed unreasonable to me to assume that a German effort would be spearheaded at Enare rather than at Rjukan.

Dean Moorland also told me that a certain amount of heavy water existed at Rjukan and that a modest production was maintained there, probably in amounts sufficient to support German experimental needs. Even without expansion, the Rjukan complex was a source of grave concern. I told Moorland of the Plough project and the plans for the destruction of power stations described in the previous chapter. It seemed safe for us to assume that Rjukan would be at the top of the list of places to receive attention.

In late September and early October, when the Plough plan fizzled out, London and Washington began to think in terms of sabotage rather than brute destruction. I had now been away from England about four months and, in mid-October, when tests showed that a superior snow vehicle had been developed and mass production could begin, I was free to go back to my Upper-Air Branch. Much fine work had been done in my absence, and I felt relieved to be back to what I liked to call "ordinary work." Three weeks after my return Sir Nelson asked me if I would go to an air base near Wick in Scotland to assist in an important operation on Norway. Much to my surprise, I found that the objective was Vemork, a major power station in the Rjukan complex, the producer and custodian of heavy water. The code name of the operation was Freshman.

At the base I met the officer in charge, a group captain, seemingly a kindly person, above the average age for his rank. I was told that it was to be a night operation and that moonlight was essential. The objective was to destroy the supply of heavy water and such installations as were necessary for continued production. Since Vemork is situated in a deep and narrow valley, the duration of moonlight was much restricted. The ephemeris showed that the requirements would be met during the period November 18 to 26, with the next suitable period just before Christmas.

The group captain, though friendly and considerate, obviously had little experience in how to obtain and use meteorological advice. On and off, he dropped into the weather station and asked casual questions, without giving away much information on the structure of the operation and such limiting factors as would assist me in framing my advice. There were no briefings or discussions to bring me into the picture, nothing such as I had experienced at Lossiemouth before the attacks on the *Tirpitz*. However, I was told that it was to be a glider operation with two tug planes. Since I knew nothing about how gliders are towed, I saw no particular problem, except such as are normally associated with customary air operations.

My attitude changed when I discovered that the glider pilot must see the tug all the time so as to keep above its turbulent wake. If the glider gets into the wake it may rip up the tail of the tug, and nothing but disaster can follow. Thus, absence of clouds must be taken as a sine qua non for towing operations.

Absence of clouds was, of course, equally necessary for the successful identification of the target. In daylight the planes could be expected to find their way through clear air, but at night things are not so simple. We had very few weather reports from the Scandinavian area, but judging by the general airflow, a cover of more or less continuous clouds could be expected over southern Norway. The weather situation across the North Sea seemed to me far from ideal. In response to my forebodings the group captain told me that he had "postcards"—messages from underground agents, presumably by radio—indicating clear sky over Rjukan.

In spite of this piece of encouraging information I felt uneasy. In mountainous terrain the cloud cover is often very variable. The postcard message did not seem consistent with the general situation. My doubts grew and my attitude hardened. In the early afternoon I informed the group captain that I considered the weather situation to be unsuitable. Apparently relying on the postcard message, he said that the conditions did not seem "too bad" and decided to launch the operation. I felt vexed at the decision and helpless to do anything about it. Throughout I had been assuming, erroneously, that the gliders were to carry Norwegian crews and I always felt that I had special obligations toward my compatriots. On the spur of the moment I said something that, in tone and substance, made it clear that I thought that cloud conditions better than those I had forecast would be required to achieve success. If the group captain had told me to mind my own business I would not have been surprised, for I had clearly slipped beyond what is appropriate for a meteorological adviser. However, he responded most gracefully and just said that as the winter progressed the weather might get even worse.

About 6:00 P.M. on November 19 the planes took off. The next morning, as I was having breakfast, the group captain told me that the operation had been a complete failure; both tugs had encountered clouds and none of the gliders had reached the target. He added that he knew his decision to go was rather against my advice and that his report on the operation would make this clear. To be "mentioned in dispatches" is a military honor in Great Britain; not to be mentioned in reports is not without its value.

Yet, I did not feel blameless. Had I known more about towing gliders I could have given more definite advice early in the day. By the time I began to understand the hazards and came to a firm conclusion, the officer in charge had, perhaps, already made up his mind. Thirty-four men were lost that night, quite unnecessarily. The heavy water at Vemork remained a worry until well into 1943 when a Norwegian sabotage group did a neat job of destruction.

The contours of the Freshman operation remained obscure to me until thirty years later when, more or less by chance, I came across a detailed account by Sergeant D.F. Cooper.[65] Apparently, the possibility of using Norwegian saboteurs to achieve the objectives had been considered and ruled out. The next plan was to drop British parachute troops, but the final decision was to land a small glider-borne force. Norwegian agents were to mark landing fields, guide the force to the target, send weather reports before takeoff, and otherwise render assistance. When the mission had been completed, the force was to break up into small groups and make their way to the Swedish border, some 150 miles away.

At about midnight signals were received from both tug planes. One asked for a bearing to bring it back to base; the second reported that its glider had been released over the sea. Careful checks of bearings showed the position of the release to be on the southwest coast of Norway near Egersund, some 100 miles from the target. This second aircraft returned to base without its glider, in the early hours of November 20. The other plane and both gliders crashed on land. Operation Freshman was over and the Nazis reported a victory. On November 22 their radio announced that a British sabotage force had landed: "the troops had been put to battle and wiped out to the last man."

On November 19, when the operation was launched, I had taken it for granted that the group captain in charge of the operation was not a member of the force. This belief was reinforced when he came to see me next morning. But from Sergeant Cooper's article I learned that the group captain flew in the tug plane that managed to return to base. When deciding to launch Operation Freshman he knew that he was risking his own life.

After the war's end a ghastly picture emerged of what the Nazis had called a "battle." Few of the men who crashed escaped injuries and all were captured. *Der Führer* had recently reinforced his orders concerning the treatment of saboteurs. The orders were carried out in full measure: torture ending in murder. Four of the men in the glider that crashed on the coast suffered severe injuries. They were taken to Stavanger Hospital and then to Gestapo headquarters for further treatment. Eventually a German medical officer ended their ebbing lives by poison. To obliterate all evidence of the torture they had undergone, the bodies were sunk, with heavy stones, about one hour's sail from the coast. The bodies of the other thirty men lost in the Freshman operation were recovered and buried with full honors.

No operation with which I was ever connected made a deeper impression on me than Freshman. Though my forecast turned out to be right, the haunting question remains with me: Should I, or could I, have been firmer in my advice early on November 19, 1942?

Chapter 14

Algiers, Tunis, and Bari

He travels best who knows when to return.
—Thomas Middleton, *The Phoenix,* IV, ii

Some time in November 1943 Sir Nelson told me over the telephone that it had been suggested that I be released from my work at Dunstable to serve as chief of a joint American–British forecasting service for the Mediterranean theater of war. He told me that the Americans and the British maintained separate services there, and it was felt that coordination between them was not as close as one might wish. I got the impression that his message reflected an American initiative. Though he did not seem opposed to the idea, he ended by saying that he hoped I would choose to stay at Dunstable.

Sir Nelson knew that my two daughters, aged thirteen and fifteen, were in Norway. Messages through the underground had kept me well informed; Liv, the younger, had developed an illness that the doctors had been unable to identify, except that arthritis was part of the trouble. The most worrying aspect was a progressive inability to digest food of any kind. While traveling through Buenos Aires in 1942 I tried to telephone, but the call was not authorized by Berlin. Friends had helped me to arrange for periodic shipments of sardines from Portugal, but very few reached their destination. Other friends in Denmark and Sweden were sending Red Cross food parcels with contents that the doctors thought would be helpful. So, in Liv's case, food was really no problem; nevertheless, her health continued to decline.[66] The doctors held out but little hope. Without hesitation I told Sir Nelson that my sole desire was to be in England when the war ended and to return to Norway as soon as at all possible.

Later in the day, reviewing my conversations with Sir Nelson, I felt that the problem was an intriguing one: Had the initiative come from the chief meteorological officers in the Mediterranean theater, or from their military commanders, or from well-meaning planners in Washington or in London? Undoubtedly, Sir Nelson would have given me the background information had I indicated any interest, but, on the spur of the moment, my reaction was entirely negative.

Early in January 1944, Sir Nelson telephoned again, saying that he thought it would be politic if, as a compromise, I could spend a few weeks at Bari, Italy, where the main British and American weather centers were located; Bari was of particular interest as the headquarters of the U.S. Air Force. For reasons he did not explain, Sir Nelson said he would like me to be back in England early in February, so time was at a premium. He could get a seat on the plane to Marrakesh tomorrow night; could I make it? My answer was yes, but I had to bore him with a trivial matter. I had only a couple of pounds in my pocket, and the banks were closed for the weekend. That would be no problem, he said; his chief meteorological officer, Group Captain Meade, would meet me at Algiers and accompany me on the rest of the tour. He would find money somewhere.

Sir Nelson sent a signal to Meade informing him of my arrival by BOAC at such-and-such a time. The girl in the Air Ministry who sent the telegram, using her common sense, decided that "BOAC" was a misspelling of BOAT. Group Captain Meade, being tied up with top-level business at his headquarters near Tunis, sent a squadron leader to meet me at Algiers. Since a large convoy was about to enter the harbor, the boat message seemed plausible, at least to an airman. There was no end to the number of boats that kept on trickling into the very generous harbor, but no sign of the expected visitor. However, since the only thing an officer can do is obey orders, the windy end of a pier was the best observation post.

In the meantime I landed at Algiers airport in the late afternoon—no sign of Captain Meade, or anyone else with an interest in my presence. After an hour or so, I tried to sort out my problem. While I was making inquiries, my coat was stolen. Fortunately, my luggage consisted of only two pieces, a briefcase and a small bag. Being adequately equipped with hands, I took no further chances. Stealing was rampant in North Africa at this time, and the word "honesty" had a meaning only in a relativistic sense. It was here that I met a British officer whose experience was of a somewhat different kind. He often left his luggage and when he returned he found someone guarding it. The guards were autograph hunters, hoping for the signature of a very famous man. The officer's name was H. G. Wells.

Eventually, I found a young officer who very kindly offered to drive me to a hotel which he, quite correctly, described as suitable for a senior officer. Like all the hotels, it was crowded, and a single room was out of the question. One thing I happen to have in common with General Ulysses S. Grant is that I do not like to undress in front of anyone. Nevertheless, I was glad to find somewhere to spend the night. Having by now only a few coppers in my pocket, I hoped I would not have to submit to a means test before I signed in at this expensive-looking place.

My roommate was not there, but in the wardrobe I found two greatcoats and caps bearing Norwegian insignia: a colonel and a lieutenant colonel. From this I concluded (as I am sure Dr. Watson would have done) that one of my compatriots was not staying overnight. The clue was that there were only two beds in the room and one of them had been assigned to me. In arriving at this conclusion I made no allowance for the possibility of the kind of "bed and boy" arrangement so well recommended by John Maynard Keynes[67] and supposed to be widespread in certain parts of North Africa. I assumed that my companions were sturdy Norwegians, like myself.

It was now 7:00 P.M. and I went on to conclude that my compatriots were having their dinner in the hotel dining room. As I entered I saw, to my delight, my good friend Lt. Col. Ebbesen and an important-looking colonel whom I knew by sight only. Ebbesen, coming from a family with long military traditions, had commenced his career in the army and had later gone into the more lucrative business of shipping. He had escaped the Nazis and was now serving the Norwegian government in exile. He was a delightful person with a keen sense of wit, highly successful as a diplomatist. His mental makeup was more typical of an artist than of an army officer. The full and important colonel had been sent on a peculiar mission: to study and report on the war effort in Africa. Perhaps erroneously, I interpreted this to mean that someone in the Norwegian government had found it useful to post him to a distant place. Then, suddenly, it occurred to me that I might be in the same boat. My mission seemed extremely vague; its need was far from obvious to me. When I told my compatriots of my plight, Ebbesen offered to pay for my extravagances; we could square accounts when I returned to London.

Since Algiers was bristling with ships and seamen in uniform I thought it likely that the Royal Navy had a meteorological unit there. Next morning I found my way to the hub of it all, the headquarters of Admiral Sir John A. Cunningham, commander in chief, Allied Naval Forces, Mediterranean. When I reached the meteorological section I found, to my immense surprise, Commander E. R. Trendell, who was now the chief meteorological officer on the admiral's staff. In the years before the war, I used to give a three-month course in weather analysis and forecasting at Oslo University to supplement the regular theoretical courses. Although it was not convenient to be absent from my Bergen home and office for such long periods, I found teaching very inspiring and profitable in more than one sense. In the first place it helped to attract new talent to the meteorological service; secondly, it was an effective exercise in self-discipline. The Royal Navy had become interested in my courses and, in 1939, Trendell was sent to Oslo to attend and report. At this time I was working on the early chapters of my textbook, *Weather Analysis and Forecasting*. Commander and Mrs. Trendell were kind enough to read the manuscript. Through their early help I felt I was off to a good start. Trendell was a keen observer and a skilled photographer. I still treasure an album with spectacular pictures of clouds, thunderstorms, and rime-laden trees which he gave me when we parted.

Trendell had developed swearing to a real art; his vocabulary was not impressive, but the variety of uses was enormous. Also, his standard response in any situation was to wave his arms and say "No problem! No problem!" When I told him of my plight: no money, no escort, and no clear idea of my destination and mission, he reeled off his standard answer, interspersed with other words reflecting his views on certain persons, all outside the Royal Navy. It was clear that his favors did not extend to the Air Ministry, including the Meteorological Office. Trendell rang up some RAF office in Algiers and was told that Group Captain Meade's representative expected me to arrive in one of the ships of the convoy. A short while before, the officer had returned to pier number so-and-so. We drove there and found an unhappy squadron leader. When we discovered that the telegram said BOAT instead of BOAC, we all laughed, Trendell immoderately. "Air Ministry!!!" was all he said when he recovered.

While Trendell was showing me some of the sights of Algiers, the admiral sent for him. The meteorological officer on duty, knowing the "best policy" proverb, explained that a certain Doctor Petterssen had arrived and the commander was helping him find his RAF contact. Pasteur is reputed to have said that "chance favors the prepared mind." The admiral had read my *Introduction to Meteorology* and he may have tried to read, or had at least seen, my textbook on weather analysis and forecasting. So he connected my name with certain problems that just then weighed on his mind. It was through this chance occurrence that I was later invited to see the admiral in Naples in connection with the Anzio operation, and also to meet him in Oslo in 1947 when he was the guest of King Haakon. Pasteur was right only in a statistical sense.

The very helpful and companionable squadron leader took me to Captain Meade's headquarters near Tunis where I was treated to an excellent briefing on meteorological organizations and problems and military structures in the Mediterranean area. I knew that Meade had started out with high promise in mathematics and had chosen meteorology as his bread and butter. His theoretical background was excellent, he had a firm grasp on practical problems, and he was enviably endowed with common sense. Viewed from any angle, he was a handsome man, able, easygoing, and diplomatic. Surely, Meade did not stand in the way of collaboration between the American and British services.

Meade's deputy, Squadron Leader Kirk, took me sightseeing for the better part of the following day, touring the countryside, looking at old ruins, and talking shop. Kirk was very knowledgeable about the history of Tunisia. It was pleasant to let my thoughts glide over the many changes, with their ups and downs, that had taken place since the establishment of the early Phoenician colonies. It seemed to me that the universus bonum of it all was very small indeed.

In my quarters, I was served by a most unusual batman: efficient, helpful, and agreeably informal. He helped to look after Churchill when he had stopped off in Tunisia a few weeks earlier before moving on to convalesce at Marrakesh. The batman was a keen photographer and very kindly let me choose three of his close-ups of Churchill, all without the everlasting cigar. Unfortunately, the pictures were lost four years later in one of my many moves.

In the late afternoon Meade and I flew on to Bari. Next morning, after an early breakfast, we went to see the forecasting centers. While Meade was transacting business with his men I called on my American colleagues. It was easy to see that something unusual was afoot. The night before a squadron of about ten planes had been sent to bomb a target near Sofia and were now overdue. The sea level weather chart showed a flat, rather weak high pressure area (or anticyclone) with light winds stretching eastward across the Balkans. Having had my mind warped, or otherwise influenced, by experience gained from the forecasting of upper winds for the bombing of German industrial centers I asked, almost instinctively, whether the force had been warned of a strong westerly current aloft. Apparently the effect of the internal structure of the anticyclone had not been considered. Clearly, strong temperature contrasts existed between the continental air to the north and the warm and moist air of the Mediterranean region. With this temperature distribution, strong westerlies aloft were bound to exist. The high pressure system was a relatively shallow affair. Here was a case, just one of many, where forecasting skill was only of secondary importance. Often, the features of the lowest layer are flighty, while the grand currents aloft are relatively conservative. Just to ascertain what is there now is often a useful approximation to what will be there in the near future. Facts are often very stubborn.

With the help of two young and keen officers we soon constructed a chart showing the topography of the 500-millibar surface. This is a surface of constant atmospheric pressure, and of variable height, broadly between 17,500 and 18,000 feet above sea level, near the level at which bombers normally fly. Routing computations based on this chart showed winds in the vicinity of seventy knots. Although upper-air soundings over southeastern Europe and the eastern Mediterranean were few and far between, there was no doubt that strong westerly winds prevailed at the level of the bombing force. What had happened was broadly this. The force had gone east with a strong tailwind, which took them far beyond the target. When this was discovered, the crews dropped their cargoes to lighten the planes and turned homeward. Having used too much fuel going east they simply had not enough left for the homeward journey.

Much of my time at Bari was spent in the U.S. Weather Center trying, when routing work permitted, to teach upper-air techniques which had long ago become standard procedure in the Upper-Air Branch at Dunstable. While doing this I had time to reflect on problems related to the training of personnel and the management of field services. An immediate conclusion was that the American Navy and the Royal Navy were more conscious of the need for disseminating reports and technical guidance to their field units than were the air forces. I was disturbed to find how slowly new meteorological knowledge trickled out to the field. As one dissatisfied officer put it, "All that happens when report (say) number 70 is issued is that it becomes filed next to number 69, usually in the office of the person responsible for the administration of the particular branch of the service."

Soon after my arrival Colonel Miller, Meade's opposite number in the U.S. Army Air Corps, came to Bari. I soon found him to be a friendly person, a keen and energetic officer, and, like most senior weather officers in the U.S. Air Force, a pilot with scant knowledge of the fundamentals of meteorology. Miller was indefatigable in visiting his field posts and extremely effective as a morale builder.

Since Meade and Miller seemed to get along very well, it was difficult at first to see why they should not pool their resources to free manpower for indoctrination of newcomers and to conduct local forecast studies and other programs conducive to growth and improvement. However, the difficulties seemed to stem not from the local chiefs but from the mother services. In 1944, traditions of the Meteorological Office spanned a period of ninety years, during which it had enjoyed a leading position in the world. Even after about 1935, when the Admiralty withdrew their interest and established a specialized service of their own, the Meteorological Office continued as a central organization serving all civilian needs as well as the army and the RAF. But from about 1920, with the advent of aviation, there had been overemphasis on routines and a marked decline from the high scientific standards of earlier years.

On the other hand, many of the meteorologists active in the war years had served in various parts of the British Empire and possessed wide experience. Although the Meteorological Office was short of trained and experienced men to meet the many and expanding needs of the present war, the number was sufficiently large to provide a thin spread of experienced men and a few first-class scientists where they were most urgently needed. At the outbreak of the war the Meteorological Office had about 800 employees of all categories; in 1944 it had over 5,000, and some of the newcomers were fully educated in the basic sciences. Though these men were lacking in experience, it was easy to see that new torches were beginning to shine.

On the American side the situation was quite different. The national weather service came into being in 1870 as a branch of the U.S. Army Signal Corps. Two decades later it was renamed the Weather Bureau and transferred to the Department of Agriculture. Other than General Albert Myer's early efforts to publish international observations, it developed few interests outside North America.[68] Under the long reign of one director, Charles F. Marvin (1913–1934), who spent much of his time in his basement instrument shop, scientific standards declined and the routines hardened. In response to new demands and in the face of rapidly expanding aeronautical developments, the U.S. Navy began to develop a small and efficient service for its own uses. However, career policies did not encourage weather officers to remain as professionals; command of ships interspersed with periods of shore duty was the only road to respectable rank. As a result, the potential that was developed by the navy in the thirties was never fully exploited.

The U.S. Army Air Corps was relatively slow in recognizing the importance of weather forecasting. When the war broke out there were only about thirty weather officers, headed by a captain; all were pilots and could not remain in "weather" and expect much promotion. Then, suddenly, in the spring of 1940, there was an awakening. Crash training programs for weather officers were started on an enormous scale and kept in a state of expansion until the war ended.[69] In the meantime, the boxes in the organization charts were filled by the promotion of young, rather hastily prepared, and inexperienced men.

During the war years the U.S. Weather Bureau maintained a business-as-usual attitude, apparently expecting that things would return to normal when the war ended. Unlike the British Meteorological Office, the Bureau played no major part in the war effort—except by sitting on innumerable Washington committees. Had the U.S. Weather Bureau occupied a position of leadership comparable to that of the Meteorological Office, there would have been a logical as well as a real basis for joint operations in the military theaters. But, in the circumstances, the basis for such cooperation was very weak indeed.

In the meantime, the Army Air Corps captains of 1939–40 had become full colonels in 1943–44. The overall leadership of the U.S. Air Weather Service changed frequently, through a strange game of musical chairs in which intrigues were not unknown. Also, there were back stairs right up to the office of General H. H. Arnold, the Commanding General of the U.S. Army Air Corps. A grunt from there sufficed to send colonels scurrying for shelter. It was only at the end of the war, when the young and brilliant Donald Yates was promoted to general officer's rank and took command, that stability and progress came to be felt throughout the service. However, in January 1944, it seemed that the history and habits of the mother services on both sides were too dissimilar to encourage hope of short-term gains by joining forces. I was glad that I had not agreed to serve as head of a joint forecasting service for the Mediterranean theater. No one can serve two dissimilar masters.

While at Bari I attended General Born's staff meetings. From the discussions it became clear that the U.S. bombers would soon be busy. The Nazis held four army divisions in reserve in the Po–Trieste area, some 400 miles to the north of Bari. It would take Field Marshal Kesselring about four days to move them down to his front, the Gustav Line. The objective was to prevent or slow down movement by bombing railways and bridges rather than destroying the reserves. Although some of the details of the plan were rather blurred, it was reasonably clear to me that the Allied Command was preparing a surprise move. From Born's manner and addresses at the staff meetings one might have concluded that he was a ten-star general with five stars on each shoulder! Actually, he was only a brigadier general. I concluded that "to bomb or not to bomb" was a question that would be decided on a higher level. Anyway, it was not my job to advise him directly; any suggestions I might have should be addressed to his chief meteorological officer. Nevertheless, the general asked me to keep an eye on the "bombing weather" for some days ahead and let him know.

CHAPTER 15

Naples and Anzio

Nature, the vicaire of the almyghty lorde.
—Chaucer[70]

My work at Bari was interrupted by a message from Admiral Sir John Cunningham inviting me to call on him on January 15 at his temporary headquarters in Naples. Taking no chances, I flew in a day early. Colonel Miller was kind enough to take me in his plane. Miller was a Texan and a patriot; aviation was his life. On the way west I talked him into taking me sightseeing over the Eighth Army front. I saw at once that my ideas of land warfare were outmoded. In the Norwegian Army at the time of the Kaiser War, infantry faced infantry, and the artillery operated from relative sanctuary, far behind the lines. Then, the infantryman was the hero; now he seemed fairly well protected, at least most of the time. While waiting for my appointment with the admiral, Miller found a sergeant to drive us up the slope of Vesuvius. As we stopped to view Naples and the Bay, I said to Miller, "Wouldn't this be a nice place for retirement?" With his typical boyish smile, Miller answered, "Texas is good enough for me." Soon after I returned to England I heard that Colonel Miller, as usual at the controls of his unarmed plane, had ventured over the Eighth Army front and had been shot down and killed by a *Luftwaffe* fighter.

On January 15, at the appointed stroke of the clock, Chief Meteorological Officer Trendell took me to the War Room in Naples, introduced me to the admiral, and then left us alone. Sir John, a smallish man with slightly stooping shoulders and the eyes of a falcon, began by saying that he had read my books and had heard a great deal about me; then he added, "But do you think your theories hold in the Mediterranean?" I recalled a conversation with General Mario Infante, director of the Italian Meteorological Service before the outbreak of the war, in which I had characterized the Mediterranean as the most complex weather region in the Northern Hemisphere. Somehow, however, an answer along such lines did not seem to fit in with the tone of the admiral's question. Being at a loss for an acceptable response, I said, with a tinge of humility, "Well, sir, mathematics and physics apply everywhere." If the answer was not a good one it could, at least, do no harm. However, without hesitation, the admiral said, "No, that is not so! Here, in the Mediterranean, God has a finger in the pie." I had not the faintest idea what he was driving at but since it was clear that he was not trying to pull my leg, it was wise for me to proceed with caution. "Well, God has always been kind to mathematicians and physicists," was all I could think of, as a harmless rejoinder. But I did not see how I could invent a third riposte if the need should arise.

I soon found that Sir John was a deeply religious man and I was glad that I had said nothing that could possibly hurt his feelings. I was glad, also, that some of my university colleagues had not heard what I said. Most theoreticians have considered me a practical meteorologist; some of the practitioners have thought of me as a theoretical man with some practical skill. Most classifications are difficult to apply to odd cases, but the truth is that my sole ambition has always been to build bridges between theory and practice, a builder of bridges—nothing more, nor anything less. Anyway, what else could I say to this remarkable admiral? We sat down and talked about the weather. His knowledge of storms on the high seas and in the British Isles was truly impressive, his questions were sharp, and facile answers would have done nothing but harm. I felt that I was off to a good start at gaining his confidence; and the gaining of the commander's confidence is the forecaster's key to usefulness. The *Tirpitz* case was very much in my mind.

The admiral explained that he found winter storms "understandable" on the high seas but not so in the Mediterranean. In response, I explained that the Mediterranean in winter acts as an enormous heat source while the surrounding continents, especially to the north, serve as a cold source: the boiler–condenser arrangement of the steam engine. In addition, the earth's rotation causes the resulting circulations to be mainly horizontal rather than vertical. These are the general principles, the *algebra* of major storms. When we come to specifics, the *arithmetic*, we must take into account the effect of topography. It is in this respect that the Mediterranean stands out as a unique weather region. The very complex system of mountains, especially on the European side, introduces modifications that, at times, are so large that they tend to obscure the general mechanisms. I obtained the impression that he saw nothing wrong in what I had said.

Next, the admiral put me into the military picture. A surprise landing was to be made behind Kesselring's front, on the Anzio beaches, on or as soon as possible after January 22. Four days of calm or light winds were needed to land the force, bring in supplies, and consolidate positions. In addition, as I had learned in Bari, railways and bridges in the north had to be bombed to prevent German reinforcements from reaching the Anzio area. The bombing would be most effective if it were done close to the time of the landing. Surprise was an overriding consideration; early bombing might sound an alert. That was all.

Commander Trendell had prepared well and his weather unit at Naples was extremely efficient. All routines were handled with laudable speed and accuracy. It was pleasant to be working with the Royal Navy again. Trendell and I soon agreed that the general weather situation was very favorable for both the landing and the bombing. A large high pressure area (or anticyclone) over southwestern Europe and the adjoining Atlantic was well banked by a series of low pressure systems (or cyclones) moving in from the Atlantic toward Norway. In such situations there is always a possibility that deep polar air may break away from the Greenland–Iceland region, move through western Europe, and set off a major storm development in the western Mediterranean. Much depends on the internal structure of the anticyclone. Judged from customary sea level weather charts, the anticyclone was of moderate intensity. But the upper-air charts showed something else. The highest level for which upper-air analyses were available was at 300 millibars, or about 30,000 feet. At this level the intensity of the circulation was greater than at any lower level. The anticyclone dominated the entire troposphere.

There is no theory or formula that can say how long such a deep and powerful anticyclone will last. When, for example, a general practitioner in medicine has seen red pustules and identified a case of measles, he usually delivers himself of a prognosis. He does so on the basis of experience, his own and that of his colleagues. Likewise in meteorology, experience must be the guide. Although Trendell and I could not say when this powerful anticyclone would weaken so much that it would cease to protect the Mediterranean area, our experience indicated that it would be an effective shield for a week or so. In the meantime the deep and strong westerlies to the north would continue to steer the Atlantic storms toward Scandinavia.

Trendell and I briefed the admiral: calm or very light winds for several days at the Anzio beaches and elsewhere along the west coast, no swell or surf from distant sources, a layer of haze along the west coast sufficiently wide and dense to hinder air reconnaissance, visibility in the Trieste region to remain good for some days. The result was that the bombing could be delayed as much as desired. Although the conditions were close to ideal, an early start was recommended. Even this powerful anticyclone would not last indefinitely.

The admiral seemed very pleased and gave us ample opportunities to supplement the briefing. As usual, his questions were sharp and to the point. Though he was not used to hearing about deep anticyclones, he asked questions about the drift of cirrus, silky streaks of white clouds in the upper atmosphere. When drifting from a westerly direction, such clouds are often the harbingers of advancing storms. He took it as a good omen that such clouds had not been seen these last few days. Since winter storms are particularly frequent in the western Mediterranean, he inquired once or twice about the possibility of such developments, with consequent westerly winds and surf on the Anzio beaches. His questions about such storms seemed rather out of context with our discussions and I began to wonder what had caused him to ask. However, he did not press this point and seemed to be satisfied that the anticyclone to the north would provide adequate protection. If he had serious doubts he did not reveal them to us.

It soon became clear that weather forecasts for the Anzio operation were seeping in from other sources. Trendell, whose ability to sniff out smelly things was far superior to mine, told me that some strange forecasts for five days in advance were being received by a U.S. naval officer from some organization in the United States. These forecasts did not reach the commander in chief through regular channels but, somehow, they seemed to find their way to high levels. Sir John's questions about possible gale winds and surf on the beaches now seemed more understandable and I began to feel uncomfortable.

When Trendell and I should meet the recipient of these mysterious forecasts, I proposed that we pool our resources and discuss the weather problems with a view to arriving at an agreed forecast or, if the need should arise, qualified advice to the commander in chief. Somehow, this did not seem to be practicable. Although precise information was hard to come by, it appeared that the forecasts came from Washington, either directly, or via some U.S. Army Air Corps organization. They were supposed to be prepared under the supervision of Dr. Irving P. Krick, based on a new system of analogs which he had developed from a series of sea level weather charts for the period 1899–1939.

I knew Krick very well. In 1934 he had spent about two weeks with me in Bergen, and in 1935, as a visiting professor at California Institute of Technology, I had worked with him for a period of four months. I considered him a very able, intuitive forecaster who could rise to considerable heights if he would dig deeper into the theoretical background of weather prediction. It was clear, even at that time, that weather forecasting was about to enter a period, perhaps a prolonged one, in which theoretical approaches would become increasingly important. Admittedly, in 1935, the gap between theory and practice was very large, but a few signposts were clear. I thought then that Krick could go quite far in this field.

However, wisely or unwisely, Krick took a liking to industrial applications and offered his services first to the film industry and later to any industry, anywhere. Krick's main protector at Caltech was its president, Dr. Robert A. Millikan, who had organized U.S. weather efforts in the First World War. Millikan was a top-level science adviser and confidant of General Arnold, the commanding general of the U.S. Army Air Corps. It was widely believed that Krick, through Millikan, exerted considerable influence on meteorological policies among U.S. Army Air Corps leaders. This indirect connection was generally thought to contribute to the variable trends in the leadership of the U.S. Air Weather Service [see chapter 14].[71]

All those rumors came to me, without much delay, from friends in America. I knew that Krick, after a brief service in the U.S. Navy, had transferred to the Army Air Corps and that his long-range forecasting system had some kind of official sanction there. General Arnold saw to it that many of the senior air weather officers were sent to Krick to study his techniques. I was well aware of these links, but what puzzled me very much was that Krick's forecasts reached U.S. naval units in the field. Though certain features seemed strange, two facts stood out clearly: The landing of forces on the Anzio beaches was a naval operation under British command, and analog forecasts for the Anzio operation were being received by U.S. naval personnel involved in the operation. I began to be resentful that I had been invited into something that, to me, resembled a hornet's nest.

I had little confidence in any system of mechanical selection of analogs and I thought it unlikely that the making of these forecasts was directly supervised by Krick. In other circumstances it would not be difficult to look up the true meaning of the word "quackery" and then ignore the forecasts altogether. There is an Icelandic proverb, almost as old as *Hávamál*,[72] saying that every man prefers the smell of his own fart. But in weather forecasting one has to try to be objective. Nature is always right, and it is extremely difficult, before the event, to rule out alternative possibilities. Almost always, one is left with a choice. In this case, the success of the Anzio operation might well depend on my choice.

The situation came to a head on January 18. Trendell and I had forecast close to ideal conditions for several days, while the so-called Washington forecast indicated a storm development with westerly winds of about twenty knots and high surf on the beaches. Nothing was vague any longer; a clear choice had to be made. Much of my self-confidence was gone. "Washington" apparently foresaw a breakthrough of deep polar air from the Iceland–Greenland reservoir, while I thought that the deep and powerful anticyclone would protect us for several days. Had "Washington" been present at Naples, we could have discussed, listened, argued, and tried to find a solution. But exchange of views was out of the question; the impersonal aspect of it all was simply frightening. I felt I was facing something enormous, something really cruel, something resembling the Boyg in Ibsen's *Peer Gynt*.[73] The possibility of a breakthrough of polar air kept on haunting me, particularly when I tried to rest. However much I turned the problem over in my mind I found it impossible to lose faith in the deep anticyclone. Moreover, "Washington" had no experience of developments in the Mediterranean area. What did they know about interactions between the upper and lower regions of the atmosphere? My self-confidence returned when I was much in need of it.

The real scare came the following day, January 19. The admiral had noticed a thin veil of silky cirrus clouds and asked how I could account for this change from the fair-weather cirrus that had been present earlier. I had already noticed the change and found that this veil of cirrus drifted from the northeast or east-northeast. I took this to mean (as the upper-air charts showed) that the good old anticyclone was still going strong. The disturbances over England and Scotland would continue to be steered eastward. Trendell, lending support, told the admiral that I had recently written a paper on the steering mechanism. The admiral seemed reassured. The landing operation was scheduled for the early hours of January 22; the meteorological briefing of the "warlords" was to be held on January 20.

Some of my MIT students were serving in the Mediterranean area and one of them initiated Operation Prof. The U.S. Air Force had taken over Cook's Hotel on a mountain facing Vesuvius for use as a recreation place for weary meteorologists, of which there were many. A dinner was arranged, students flew in from various outposts, and hopes were high for a pleasant evening. Since I had to brief the admiral at 9:00 the next morning, I accepted the invitation on the condition that I could retire at midnight and that two Jeeps with drivers would be ready next morning sufficiently early to land me at the commander in chief's headquarters not later than 8:00 A.M. In case of car trouble, the following Jeep would carry on. This was accepted. Two sergeants selected as drivers were told to go to bed early.

During January 1944 Vesuvius had become rather active. On the evening of the dinner the volcano put on an extra show. Fiery props in large numbers flew high in the air, and a stream of white-glowing lava about 800 feet wide and twice that long brightened the sky—a truly beautiful sight, for which the M.C., Colonel Sorey, claimed credit. The evening was unforgettable, in the true meaning of the word. It made me realize, again, how pleasant is university life: to be free of care and to mix with young people. Next morning we all had to go back to the grim business of war.

The Jeeps ran into trouble. An army convoy messed up traffic in the lowlands and the following Jeep disappeared. My driver was fully aware of his orders and, as we approached Naples, he drove on sidewalks among screaming and angry pedestrians. Eventually he rounded the convoy and sped on to the admiral's headquarters. Although beads of perspiration were plentiful on his forehead, he was calm as a monument of Caesar: "0758, Sir!" was all he said. "Thank you, sergeant!" was the appropriate response. Without such people the world would be very dull.

At the appointed time my audience gathered. While I was standing aside waiting for the proceedings to begin, I found it amusing to paraphrase Churchill by (inaudibly) saying that never in the course of human endeavor has so much rank been condensed in such a small space to listen to so little. My audience included Admiral Sir John A. Cunningham, General Sir Mairland (Jumbo) Wilson, General Sir Harold Alexander, and a number of other officers of high rank.

The chairman, Admiral Cunningham, opened the proceedings by introducing me. He reminded his colleagues of the poor forecasts they had received for earlier landings, especially in Sicily and in southern Italy. "But" he went on, "this time I have succeeded in obtaining the services of *the* expert. . . ." While the admiral spoke I reminded myself of the analog forecasts, with gale winds, etc., on the Anzio beaches. After such an extravagant introduction by a highly respected admiral who also knew a great deal of meteorology, I felt I could not afford to make a mistake. A compromise forecast was out of the question. What I did, on the spur of the moment, was to cut out the fourth and last day in my forecast.

I explained in brief and simple terms the evolution of the general situation, the protective anticyclone to the north, the reasons for calm weather in the western Mediterranean, the stability of the haze layer along the west coast, etc. I concluded with a forecast for ideal operational weather for three days. Since a four-day forecast was desired, I thought I could cover myself by adding, "This does not mean that the fourth day will be bad; it only means that I am unable to foresee events beyond the third day."

Alas, my stratagem did not work. General Alexander (later field marshal, earl of Tunisia, governor general of Canada, minister of defense, etc., etc.,) debonair, aristocratically easygoing, the most human and unpretentious officer I have ever met, caught me. "If I asked you privately, not with these people about, what the weather will be on the fourth day, what would you say?" he asked me with an eloquent smile. Obviously there was no point in further hedging, so I answered, "Sir, I think it will be pretty good!" Everybody seemed to be pleased. The admiral ended the session by saying that my forecasts agreed with his own. From a professional standpoint he had a perfect right to say so; his knowledge of meteorology was truly remarkable. I never felt I had persuaded him. I felt we had discussed and agreed.

The admiral asked me to check with him at 6:00 P.M. All I could do was confirm the earlier forecast. Things were going very well. He said he would like a last check at 9:15 P.M. Soon after six o'clock the admiral and his colleagues went off for dinner and, I imagine, a small drink. But at the appointed stroke of the clock the admiral was back, checking on details and receiving the latest reports from members of his staff. All day he had been checking and checking; I could only admire a man who could exercise such control without becoming submerged in detail. Then, in a rather relaxed tone, he asked me if the weather would hold. My answer was as firm as before: "Sir, I have nothing to add or subtract;

there will be at least four days of fine weather." Then, I witnessed something that moved me deeply: The admiral turned away from his crowded desk and prayed for success. It was pleasant to feel that the word "soul" stood for something very real. I felt honored that I had been allowed to be present.

History records that the landing at Anzio came as a complete surprise, the weather held well beyond the fourth day, men and equipment were landed and forces built up as planned, and the naval phase came to an end. In the meantime, it was pleasant to relax and to recall what little I knew of Roman history and, especially, the glory of ancient Antium, the intrigues of Agrippina, and the extravagances of Caligula and Nero. Surely, there was but little left for Anzio to boast of as the war was drawing to an end.

It was instructive to observe life in Naples and the degrading as well as other influences of the war. Young lads were seen to act as procurers and their young sisters or friends would produce the merchandise; the oldest profession was the first to recover. Hardworking people who had lost their livelihood were trying to reestablish themselves in business. Money—even in small amounts—was needed to get back on the road. Small antiques and art shops were beginning to fill and American servicemen were curious and had money to spend. Allied bombing had left its marks, and so had the deliberate burning of houses by the Germans to destroy facilities that might be useful to the Allies. None of the hotels along the waterfront were spared. A room had been reserved for me in the Vittoria, which looked as if it had seen much better days. It had been burned out, and restored to house senior officers. Being something in the nature of a special guest, I had a room with improvised running water. When the boards from its glassless windows were removed, I had a grand view of the Naples harbor and the masses of ships that gathered there before the Anzio landing. The floors of the dining room and the lounge, on the ground floor, consisted of sand, deep, clean sand, taken from nearby beaches—quite as attractive as most modern carpeting. Food was not too much of a problem and the service was excellent; smiling waitresses, anxious to please, go a long way to make one forget material shortcomings.

The owner of the Vittoria was a very interesting person. He was a man in his early forties, well educated, handsome, and obviously used to wealth. When the war broke out he owned several hotels on the waterfront, a house in the city, and a villa up in the hills. Early in the war the government requisitioned his Mercedes-Benz. When the Germans arrived in force, his second car was exchanged for a slip of paper. He then went on the black market, was eventually found out and fined, and the car was removed without much paperwork. A second search of the black market was successful and, for security reasons, the car was hidden, most of the time, in the garage of his villa. During a British raid a bomber overshot the target and destroyed the garage and the car and, incidentally, removed the villa of his neighbor. After a while, he acquired his last car and paid heavily for it. But now he had no garage and stealing was rampant. So, as a security measure, he bought a heavy iron chain and a bulky padlock and chained the right rear wheel to a lamppost. Next morning he found the lamppost, the chain, the padlock and the right rear wheel—all in good order. But the car was gone! The spare wheel had served its purpose.

As he told his lengthy story in somewhat *gebrochen* English, he paused at the end of each little tragedy, slapped his thigh, and roared with laughter—roared as if Apuleius himself had told his tidbits from "The Golden Ass." He was a marvelous person. How I wished that my own temperament had been more akin to his. On my last night in Vittoria he sat down with me for dinner and brought with him a bottle of old wine.

I met Sir John Cunningham again in 1947. On two historic occasions he had been associated with King Haakon, first in 1906 when he, as Prince Carl, became King of Norway, and again in 1940, when the king and government escaped the Nazi onslaught and went into exile in England. At the end of the war Sir John served as first sea lord. It was not until 1947 that he had time to avail himself of an old invitation to come to Oslo as the king's guest. During his visit the Anchorite Club (of which I had never heard) offered a luncheon. I was invited, not only to attend but also to make the first speech. I was told beforehand that everything would be on the informal level. This is what I said:

> Your Majesty,
>
> I appreciate the honor of being called upon to say a few words on this occasion, and I'll show my appreciation by being very brief. On January 12, 1944, I received a telegram that gave me much surprise. It was a signal from Sir John Cunningham asking me to see him at his headquarters in Naples.

> Naturally I was a little nervous when I had to appear before one of the great warlords, but I was immediately put at ease. We at once drifted into a conversation on weather forecasting in the Mediterranean. It was a learned discussion, and I am glad to admit that I picked up some useful points from it.
>
> At this time I had just written a paper on some—as I thought—rather intricate upper-air conditions that influence the development of storms. Much to my surprise, Sir John knew all about it; he had not read my paper, but he had made his own observations and drawn his own deductions.
>
> It was very interesting to me to watch everything that went on at Sir John's headquarters: the hard work, the magnitude of it all, the precision and care with which everything was planned, and the admiration for the commander in chief that radiated from all members of his staff. I believe I radiated as much as anyone else.
>
> In the evening of January 20, when everything was ready for the landing at Anzio, I left the headquarters with a feeling that nothing could go wrong. Nevertheless, I sat up until quite late, smoking many cigarettes, and wishing success to Sir John and his navy. Nothing went wrong; nothing could go wrong; nevertheless it was pleasant to wish.
>
> The British Navy has in the past defended more than Great Britain. It has been a stabilizing factor of great importance to all seafaring nations—and, as we have recently seen, to many landlubbing ones as well! Wars end but problems remain, and the Royal Navy will remain.
>
> Your Majesty, may I propose that we raise our glasses and wish success to the first sea lord and the navy "in which he serves."

Sir John responded with a rare mixture of wit and grace, and proposed a toast for the Royal Norwegian Navy, after which the king, speaking as "the senior Norwegian naval officer present," followed. With the simplicity of wisdom so typical of him, he expressed what we all felt deeply: the strong bonds that have long existed between our two seafaring countries, Great Britain and Norway. While sitting at the king's table I found it amusing to remind myself that I might not have been there at all had not a darling girl in the British Air Ministry, now so far away, changed BOAC to BOAT.

Soon after my return to Bari I received a message from Sir Nelson asking me to be available in London on February 2. Without assuming very much I took it that the business was connected with arrangements for the assault on Nazi-held Europe, which I had thought would begin in May. Two days later I received, this time through American channels, a message from my good friend Rossby saying that he and Colonel Senter were in London and would visit North Africa and Italy. He would like to see me, preferably in Marrakesh. I had not seen Rossby since March the previous year and I was overjoyed at this opportunity of a mental romance, just when new things were beginning to take shape.

However, on further consideration, it seemed that certain items added up nicely while others did not. In the first place, Rossby was one of a few science advisers to the secretary of war, Henry Stimson. Rossby's strength derived from his flair, vision, and drive rather than from his judgment in the making of practical arrangements. On the other hand, Senter was a senior colonel (who later rose to high rank), known for his cool judgment, thorough staff work, and diplomatic skill. Colonel Senter would not be traveling with Rossby unless something big was in the offing. Most likely, the business was the same as that which had prompted Sir Nelson's message.

Second, there were rumors about. General Arnold had shown renewed interest in the management of the U.S. Army Air Weather Service. Krick had been, or was soon to be, posted to England. He might even be assigned to General Eisenhower's staff. Krick's analog system, also called "canned memory" (or "tinned memory" in the British version of the English language), had received official blessing, etc., etc. However much I deducted to allow for the verity of rumors, I was left with a formidable residual.

Moreover, my recent experience with analog forecasts was far from encouraging. In my judgment, the meteorological aspects of the Anzio landing were quite simple. Commander Trendell could certainly have come up with the right solution without my help. My presence was helpful mainly because I used what authority I had to oppose, with reason and firmness, outside interference and suppress the worries that resulted from the "canned-memory" forecasts so irresponsibly provided by personnel without any experience in the meteorology of the Mediterranean region.

There were further considerations. Rossby had many likes and dislikes. Although, given some time, the walls between the categories were quite permeable, he had maintained an enduring dislike to Krick's methods, claims, and salesmanship. The first modern university department in meteorology in the United States was organized by Rossby at MIT. When, in 1935, Millikan established Krick at Caltech, a competitive element entered the scene. Objectively, however, no one with real knowledge of the subject could consider it a question of equality. In mathematics the statement

$$A >> B$$

means that the quantity A is very much greater than B. In my opinion it was a statement of this kind that would best describe the scientific quantities involved, though, admittedly, Krick had practical skills not possessed by Rossby. In forecasting, practical skills cannot be ignored.

As to the Krick controversy, both the opposition and the following were strong and widespread. Personally, I stayed out of it until 1940. Then, at a meeting of business interests to which Rossby and I had been invited, I heard Krick saying that he was using "the Petterssen methods." On this occasion I probably overreacted in dissociating myself from his claims. Had this happened in, say, 1970, I would probably not have reacted audibly. In 1940, however, I was a newcomer to the United States and unfamiliar with American advertising techniques. I was Rossby's successor at MIT, I was proud of what I headed, and I was anxious not to become associated with anything that I considered to be of doubtful value. Though I may have overreacted, some negative response was clearly called for, particularly since I was called on to express my views.

Some years after the war's end, when I reviewed past events, especially Krick's propaganda for his analog techniques, I recalled a passage in Book III of Carlyle's *Past and Present* (1843) which, when properly paraphrased and muted, rather well expressed the view I held while at Naples:

> He has not attempted to *make* better hats, as he was appointed by the Universe to do, and as with this ingenuity of his he could very probably have done; but his industry is turned to *persuade* us that he has made such.

In January 1944, however, the problem before me was a practical and not a philosophical one. It was very difficult indeed to reconcile the rumors about Krick's posting with anything that I thought Rossby would be willing to recommend. On the other hand, and with no less force, there was the argument that what Rossby liked most of all was influence. I could not quite see how he could find it opportune to recommend anything athwart General Arnold's wishes. Then there was Krick's connection, through Millikan, to General Arnold. Rossby, though brilliant and dashing, was not reckless in matters of this magnitude. Something, both big and shapeless, seemed to be afoot, something that I did not understand.

After much "perhapsing" I thought I saw a clear road. I was working for the Meteorological Office and I had developed an attachment to Sir Nelson. The unwritten code of loyalty demanded, I thought, that I avoid exposing myself to persuasions and pressure before I had heard Sir Nelson's wishes. I was very sorry indeed to miss this opportunity of meeting my good friend Rossby, this incomparable source of inspiration and master engineer of interesting episodes. On my request, arrangements were made to advance my departure by twenty-four hours. On the night chosen, our paths—Rossby's and mine—crossed somewhere over the Bay of Biscay.

Chapter 16

Overlord: A Dream and an Awakening

Hope is generally a wrong guide,
though it is very good company by the way.
—Edward Halifax[74]

On my way back to London I had plenty of time to consider the problems that I assumed were on Sir Nelson's mind. To me the general situation did not seem particularly complex. The supreme commander, his deputy and the commanders of the assault forces would be a mixed team, consisting of American and British officers, and the forces would be overwhelmingly composed of American and British units. It seemed certain, therefore, that the group of forecasters that were to provide meteorological advice to the high command had to be a mixed American-British team; a fifty–fifty composition seemed altogether reasonable.

Neither as a Norwegian employee nor as an American immigrant was there a place for me on the team. However, since I had been invited by the British to develop their upper-air program and otherwise assist in the forecasting for important operations, I thought that Sir Nelson would wish me to join as a member of his organization. Additionally, since Sir Nelson's deputy, Mr. E. Gold, was in direct charge of all weather services, I thought that he, too, might wish me to join. There were four or five assistant directors, all elderly, and rather worn after years of work in a highly pyramidal organization, but none of them had much say in matters of policy. Although the general situation seemed simple enough, complexities were likely to develop when it came to arrangements involving people. Since I expected to see Sir Nelson immediately after my arrival in London, I thought it wise to do as much homework as possible and to be prepared for an awkward question that might arise.

I took it for granted that the head of the meteorological team had to be an experienced forecaster, a primus inter pares,[75] with outstanding skill in diplomacy. Obviously, I thought, the spokesman for the meteorological group must be able to present briefings and answer questions with convincing professionalism. In dealing with officers of exalted rank, he would not get very far without firmness dressed in the garb of

Fig. 16.1: C. K. M. Douglas, British forecaster, who, among other things, contributed greatly to the choice of D-Day.

diplomacy. This conclusion, however, led to the baring of a problem: Since no American had much experience in forecasting for western Europe, the head of the group ought to be a British meteorologist. Would the Americans agree? If my opinion should be asked I would have to speak for the best meteorological solution, regardless of national or interservice rivalries.

Among the British forecasters my colleague at Dunstable, C. K. M. Douglas (Fig. 16.1), stood in a class by himself. He had a fine background in science and about twenty-five years of experience in forecasting in Great Britain. Douglas was an intuitive forecaster. His most remarkable talent was an extraordinary memory of the evolution of past weather situations. His memory was so precise that he could, almost blindfolded, pull out of the bulky map files any weather situation that he had in mind. No meteorologist, anywhere in the world, could match Douglas in his skill at predicting fog, visibility, clouds, and rainfall in the British Isles, though some could do as well in forecasting the movement of large-scale motion systems and their evolution over extended periods.

While serving as a pilot in the First World War, Douglas had developed a nervous affliction, which was likely to rule him out of consideration for the job as head of the group of forecasters that I assumed would have to be formed. On the other hand, a team of forecasters without Douglas seemed simply unthinkable. He and I had worked side by side under heavy pressures for a period of more than two years. I liked to think that we had developed a considerable measure of mutual understanding and respect. I never argued about his estimates of clouds, weather, and visibility, and I cannot remember any occasion when he took exception to my forecasts of upper-air structures. A similar relationship existed between our staffs.

On the (as I thought) reasonable assumption that the head of the meteorological group had to have the qualifications just stated, it did not seem easy to find a suitable person within the British Meteorological Office, unless Sir Nelson was prepared to violate the time-honored principle of civil service seniority. If this were done, several highly qualified men would deserve consideration.

The timing of the initial assault on the continent of Europe seemed likely to be dominated by naval meteorological considerations. What was good enough for the navy would certainly be satisfactory for the army. The air force was always, and rightly so, demanding. Although their requirements were highly varied, none of them, nor their sum, would be comparable with the do-or-die question of the navy: Can we, or can we not, land the force in fighting shape on the coast of France?

In earlier years the Meteorological Office had provided the service that they thought necessary for naval uses. Rightly or wrongly, the Admiralty felt that their needs were not well met. In the early thirties they started divorce proceedings and established a meteorological service for the fleet and its air arm. From its inception, the Naval Meteorological Service was directed by Captain L. G. Garbett, whom I had met in Germany in 1939 [see chapter 4]. Garbett, though not a scientist and forecaster, had an extraordinary gift for attracting talent to the service, stimulating progressive schemes, and leaving able men free of red tape and administrative fetters. In November and December 1941, when I worked with the Royal Navy on problems related to Norwegian interests, I had become well acquainted with some of their leading meteorologists. Several appeared to be unusually able and well equipped. Here, too, one could find men qualified for the job as head of the team.

Since naval requirements would loom large in the vital operation of landing and supplying the assault forces on the coast of France, I took it for granted that meteorologists from the Royal Navy would be assigned to the team. Professors Walter Munk and Harald Sverdrup, at Scripps Institution of Oceanography in La Jolla, had developed methods of forecasting swell and surf with the aid of meteorological charts. These new techniques had been tested by the Naval Meteorological Service and adapted for their special needs. As far as I knew, no one else had real competence in forecasting the state of the sea in the Channel and the surf on the French beaches.

As I continued my survey, I became increasingly optimistic. The unknown factor was what the U.S. Army Air Corps would agree to. If the rumors afloat had any substance in them, Krick would be a prominent member of the team. I considered it equally certain that he would see to it that officers trained by him would be posted to the group. Furthermore, since the machinery for selection of Krick's analogs could hardly be moved to the supreme commander's headquarters, one would have to reckon with some degree of interference, with analog forecasts coming from Washington. However, in a balanced group of recognized forecasters working together as a team, the merits of the analogs would be discussed and compared with those of other methods; in these circumstances, they might well prove to be a useful handmaid to more powerful techniques.

In spite of the agonizing experience I had suffered in connection with the Anzio operation, I was not opposed to the use of analog systems as such. What I was opposed to, and rather strongly so, was the inept use of Krick's system by the Washington group. The practical problem, as I saw it, was to minimize the effect of the Washington interference and to develop a cohesive team. For this very reason, the composition of the group of forecasters at the supreme commander's headquarters seemed to be a matter of paramount importance.

I thought that the team could be greatly strengthened if the Americans or the British would invite Professor J. Bjerknes to join. Jack had discovered the polar front and authored the polar front model of cyclones. In about 1936 he was the first to describe the associated waves in the upper atmosphere and to explain their mode of propagation. With his father and many collaborators, including such luminaries as Tor

Bergeron and Erik Palmén, he laid the foundation of what is often called modern meteorology. Indeed, even today, forecasters throughout the world will agree that large parts of their mental equipment have come from the works of Vilhelm and Jack Bjerknes. I was fortunate enough to have had the opportunity to work, on and off, with V. Bjerknes, and to have been closely associated with Jack for a period of eleven years. I knew of no one who could be more useful than Jack in creating cohesion within the group if tensions should arise.

There was one further consideration. As early as 1920 Douglas had spent a few weeks in Bergen to become acquainted with Jack's polar front ideas. Many years later Jack, in response to an invitation, had worked for several months in the Meteorological Office. As a result of these and other exchanges, a close and lasting association between Douglas and Jack had developed. When the war broke out in Europe, Jack happened to be in the United States and was prevailed upon to contribute to the U.S. war effort by establishing and directing a department of meteorology at the University of California at Los Angeles. I knew he would respond unreservedly and immediately to any call to useful service against the Nazis.

After these very pleasant dreams came a rather abrupt confrontation with reality. Upon arrival in London I was received by Sir Nelson. He, with the deputy director present, explained that they had been considering arrangements for the provision of meteorological advice to the supreme commander of the assault on the Continent of Europe which, it was assumed, would take place in May or June. To ensure that normal meteorological services would not be disrupted, they had thought it best that the Upper-Air Branch and the branch headed by Douglas should continue their work at Dunstable. Similarly, the U.S. Air Weather Service and the forecasting center of the Royal Navy would continue their normal work. It became customary to refer to these organizations as Dunstable, Widewing, and the Admiralty. These notations will be used in what follows.

The three centers would be connected by scrambled telephone lines. Each would, as an additional duty, contribute to the forecasting for the operation, which later became known as Overlord. All that was needed at the supreme commander's headquarters was a senior meteorological officer to serve as the chairman of telephone conferences between the participating centers. He would smooth out differences of opinion (should such occur) and formulate the advice to be given to the supreme commander. As I listened, the word "straw" flashed through my mind: a man of straw, catch at a straw, to make bricks without straw, etc. Perhaps it was flashes of this kind that Virginia Woolf had wished to allude to when she wrote about "voices that fly ahead." *Straw!*

Before I could find some helpful response, the deputy director explained that the arrangement would be similar to the one that had worked so well in providing advice to the chief of Bomber Command. In my opinion the similarity existed on paper only. Each bomber group had its own forecast unit and there was a fully equipped and well-staffed center at Bomber Command headquarters. The chief meteorological officer at each of these centers was a hand-picked, experienced forecaster. The telephone conferences with Dunstable, which were entirely consultative, were chaired by the meteorological officer in charge at Bomber Command headquarters. He, and he alone, was responsible for the advice given to the chief of Bomber Command. His advice was based not only on the conference discussion but also on the fine work in his own forecast center. The purpose of the telephone conferences was to achieve general coordination, disseminate views, and, especially, keep the centers posted on the thinking of the two branches headed by Douglas and myself.

During these bombing conferences we sometimes heard faint clicks in the microphones, as if someone (within the scrambled system) was plugging himself in or withdrawing. We knew that air marshals at group or command headquarters often listened in on our discussions, but this prevented none of us from expressing frank opinions about possible weather developments. Why else were we on the telephone lines? Surely there were disagreements, but, in each case, as the conference progressed, the opinions converged. I cannot remember a single occasion when the chairman's summary did not reflect substantial agreement—though, on occasion, residual differences were included also.

The bombing conferences were competently chaired and they produced important by-products: The air marshals became educated to the proper use of meteorological advice and advisers, and they learned to appreciate uncertainties and risks in cases where such were important. There is clearly a difference between a blind risk and a calculated risk. The meteorological conferences served to reduce the blind risks to a minimum. Toward the end of the war, the chief of Bomber Command said that his job had become ninety percent meteorology and only ten percent strategy. Though this statement was not meant to be factual, there is much truth in it. I doubt that he would have said so when the bombing began. The night bombing and the meteorological service for night bombing grew up together.

Why did the Bomber Command conferences work so well? In the first place, the area of interest was a rather narrow one: visibility at takeoff; wind and icing en route; cloud cover over target; and fog or no fog when returning to base. Secondly, all participants—except me—were British, and in the same general age group. They had fairly uniform background, training, and experience, and they all knew one another personally. All of us stood face to face with the same awful responsibility; all were eager to receive and give help. I, for one, was proud to be associated with this team.

The similarity between the bombing arrangement and the one now contemplated for the Overlord operation was, I thought, largely limited to the use of telephone lines; almost all other components were essentially different. In the case of Overlord, the area of meteorological interest was extremely wide: visibility, swell, wind, surf, and shelter for naval vessels; trafficability for army vehicles; and visibility, cloud amounts, and winds at different levels to meet the requirements of the whole conspectus of air operations. In the case of bombing, the time scale of the forecasts was generally twelve to eighteen hours. The question was: Can we take off, bomb, and return to base this coming night? In the case of Overlord since slow-moving surface vessels from harbors as far north as Scotland were involved, the weather over a period of two, three, or even four days had to be considered. In the case of bombing, the time between "go" and "attack" was of the order of six to twelve hours; in the Overlord operation it would most likely be thirty-six to seventy-two hours. When viewed against this background, much of the similarity vanished.

Other differences were even more alarming. The three groups—at the Admiralty, Widewing and Dunstable—were composed of men of widely different backgrounds, training, and experience. To me it seemed totally unrealistic to expect anything resembling the cohesion that had characterized the bombing conferences. Over the years I had gained some experience about team differences by working in the three forecasting centers of Norway (Oslo, Bergen, and Tromsø), in the U.S. Navy, the U.S. Weather Bureau, the Canadian Meteorological Service, at Dunstable, and, for a short while, with the Royal Navy. I had some knowledge of the forecasters in the Admiralty, and I knew Krick and some of the officers at Widewing. I found it hard to believe that good work could be done by telephone discussions among such diverse groups.

External conditions added to the difficulties. During the war, observations from the North Atlantic and from enemy-held territory were extremely sparse. It was difficult, even at Dunstable, to maintain consistent sequences of analyses of current weather situations. It was not too difficult for me to imagine the discrepancies that were almost certain to arise when three rather dissimilar teams attempted to obtain not only mutually consistent analyses of current weather situations, but also agreed-upon predictions of the evolution of weather and motion systems over extended periods of time.

Still worse, the chief meteorological adviser to the supreme commander, unlike the chief meteorological officer at Bomber Command, would have no forecast center at his disposal. He would have to rely entirely on what he heard over the telephone lines. On top of it all, the chief meteorological adviser would be Dr. J. M. Stagg, who was neither a forecaster nor a climatologist. Stagg was a meteorologist mainly in the sense that he was employed by the Meteorological Office. His scientific field was terrestrial magnetism and he had done some work in radiation, but none in or anywhere near subjects related to forecasting. He had advanced up the civil service ladder through work in the fringe areas of meteorology and in administration and management. In Great Britain civil servants normally retire at sixty. Stagg (who was born in 1900) had now reached the age when deserving men are being considered for top positions. In 1944 the deputy director was three years overdue for retirement and the director had eight more years to go. Within the Meteorological Office it was widely thought that Stagg was being groomed to fill the first vacancy.

After a brief introduction and a description of the skeletal structure of the organization contemplated, Sir Nelson invited comments. I responded with what I intended to be forceful arguments for a team of forecasters working together in one place. To reinforce my arguments I pointed out that I was not interested in rank or status within the service. I would be very glad to serve on the team and I would certainly understand that there might be good reasons for leaving me out. I would gladly continue my work at Dunstable. Since my branch was the only unit (British or American) that had specialized in upper-air work, I would be glad to make myself available for telephone consultations at any time should the team desire information or advice. On the other hand, I could not conceive of any competent organization to handle the Overlord business except a well-chosen team of forecasters working as a unit. I also stressed the advantages that would accrue if Professor Bjerknes were invited to join. Sir Nelson said he would explore with the Americans the possibility of inviting Bjerknes, but apparently this suggestion was not found to be useful. At this time some Air Council business claimed Sir Nelson's attention and he suggested that the deputy director and I continue the discussion, with a view to narrowing the gap between our different views.

The discussion that followed became rather lengthy and to some extent repetitious. Although the alternatives were either a team or no team at all, a fifty–fifty compromise did not seem attractive. Once or twice I obtained the impression that the decision had already been made, and I saw nothing to be gained by continued discussion. However, the deputy director reassured me in rather firm phrases, saying, ". . . we are not wasting our time as long as we are trying to find a solution."

I found the situation rather anomalous. Surely a solution to the problem did not have to be negotiated with me. Although I was a Norwegian employee, I was filling a position in the British service and had to take instruction from them. For my part, if I thought it necessary, I could ask Admiral Riiser-Larsen, the commander in chief of the Royal Norwegian Air Force, to recall me. I felt that I had now come under uncomfortable pressure to agree to something that I considered technically unsound and that might well jeopardize the operation. With such thoughts at the back of my mind I told the deputy director that I did not see how I could contribute usefully under the arrangement they were considering.

Sometimes it is well to refer back to first principles. To me a scientific process is one in which things become reduced to order. The purpose of management is to facilitate such reduction. The scheme proposed seemed to be widely at variance with what I considered sound principles. The chief meteorological officer at the supreme commander's headquarters could not be expected to make bricks without straw. Anyway, my opinion had been invited, and I saw no reasonable basis for yielding or compromising.

At about this time, Sir Nelson returned and, with his customary charm, asked if we had solved all problems. The deputy director, perhaps jokingly, said that I had indicated my intention to withdraw. In the circumstances I felt it right to summarize my arguments for a team of forecasters and to repeat my doubt about my own usefulness.

Sir Nelson had more business to attend to and the deputy director and I drifted into a chatty discussion of difficulties inherent in multinational arrangements. Strangely enough, his remarks now seemed to support, rather than contradict, my arguments for a team working in the same place and cultivating personal contacts. Toward the end of this conversation, he pointed out that the assault on Nazi-held Europe would be a very complex affair, with American and British forces, facilities, and services intermingled. He also indicated that *the meteorological organization was to be such that if anything should go wrong the responsibility could not be identified with any particular organization.* I was stunned by this remark, and for a long time I was unable to see ameliorating circumstances.

As I write, twenty-nine years later, I find it quite easy to see that his remark may not have been deeply meant. Perhaps the organization proposed was the only one obtainable through negotiations with the authorities involved; perhaps the diffusion of responsibility that the proposed scheme provided for was an accidental by-product of arrangements imposed by higher authorities; perhaps he felt he had to defend something in which he had but little confidence. Who knows?

But in February 1944 I saw these things in a different light. I had looked forward to the assault on Nazi-held Europe as a gateway to freedom, not only for my daughters, relatives, and friends in Norway, but for Europe as a whole. It would be the beginning of the end of senseless destruction and slaughter, of a formless horror, of the rule of a sadistic madman, and, perhaps, the dawn of something resembling what Nordahl Grieg had in mind when he wrote *Greater Wars*.

Viewing the situation against this background I felt bewildered. I was tired after a sleepless night and a day without lunch. I knew of a small Greek restaurant in the neighborhood and thought that a meal and a glass of wine would do me good. When the owner met me, as he had done before, I found that my appetite was gone. So, instead, I asked to use his telephone. I phoned Riiser-Larsen and asked to see him. Though it was late in the day he found time. I told him of my misgivings and asked what his reaction would be should I find it necessary to request withdrawal. In his memoirs Riiser-Larsen states that I asked to be recalled and that he refused my request.[76] Undoubtedly, his memory must have done him a disservice. In the first place, I had only asked him, as a friend, what his reaction might be; secondly, on his suggestion, we consulted a Norwegian official who was well versed in international matters. His advice was not to act on anything hypothetical but rather to wait until something specific had occurred that could fully justify my request.

No further information about the proposed organization was allowed me at the time. In the years since I have been able, gradually, to fill in some of the missing detail. And then, in 1971, Dr. Stagg published his account of the affair under the title *Forecast for Overlord*.[77] Stagg's book, by his own description popular and not scientific, seems to me to contain a number of inaccuracies, to misrepresent what actually happened in the early months of 1944, and to give a distorted picture of the contributions by some of the forecasters who took part in the work. I shall have occasion to refer to some of these aspects later. Here, however, it is of interest to note that Stagg's book makes it clear that the general arrangements for the provision of meteorological advice to the supreme commander were negotiated with American authorities in November 1943, leading to a principal agreement that a British meteorologist should fill the position of chief meteorological officer to the supreme commander. Thereafter, Sir Nelson nominated Dr. Stagg for the position and the appointment was effected in late November, with a U.S. Army Air Corps colonel as deputy.

We learn also that in January 1944, Dr. Rossby and Colonel Senter, representing high military authorities in Washington, inspected the operation of the services provided by Stagg. Soon after Rossby's departure, Stagg's deputy was replaced by a U.S. Army Air Corps colonel who had attended Krick's courses, while Krick was posted to the American forecast center commonly known as Widewing. These changes were followed by the demotion of Stagg and the promotion of his new deputy to the top position. After intervention by British authorities, with reference to the principal agreement, this turnover was reversed and Stagg was again in charge. Clearly, the meteorological part of Overlord was off to a bad start. Neither the many behind-the-scenes maneuverings nor these major upheavals that happened in broad daylight held out much hope of smooth teamwork.

While Rossby had stressed the importance he attached to the forecasting services for the launching of the Overlord operation, British authorities seemed to have taken a more bureaucratic view, reflecting emphasis on management rather than technical skill. Stagg has Sir

Nelson actually saying that the job of chief meteorological officer to the supreme commander

> . . . shouldn't be so arduous. In spite of pressure from Moscow, the Allies are most unlikely to be able to open a second front before late spring and by that time of the year, weather has usually quieted down from its winter storms so that forecasts from the various central offices are not likely to differ much from each other.

It is well to note that, here and elsewhere, the quotation marks in Stagg's book refer to extracts from his notebooks and need not necessarily indicate the words of the persons indicated.

It also appears from Stagg's book that he was under a serious misapprehension about my duties and the work that I was doing at Dunstable. The question of provision of forecasts for several days in advance was raised repeatedly, not only by officers on the supreme commander's staff, but also by Rossby and his associates. When I joined the Meteorological Office [see chapter 8], I had made it quite clear that my desires did not include long-range forecasting research. The British effort in this area, which was well under way when I arrived, folded up in 1943 without any indication of useful or promising results. To the best of my knowledge, no new effort was initiated during the war. At no time had I in any way, even remotely, been associated with any British effort in extended or long-range forecasting research. And yet Stagg (pp. 20–24) told Rossby that I had been invited into the Meteorological Office "to explore and exploit the use of the new upper-air information in the making of forecasts for two, three, or more days ahead." When pressed for specifics, he explained, "what was going on [in January 1944] with great urgency at Dunstable," and, on another occasion, "what Dr. Petterssen was doing [in January 1944] in his special upper-air section." These completely unfounded assertions were particularly annoying. My attempt at removing possible misunderstandings resulted in an exchange of correspondence that is reproduced in Appendix A.

It was not until late in 1947 that Rossby and I had occasion to review the Overlord business. Rossby felt then that he had been misled by Stagg's description of my duties and work at Dunstable. In fact, Rossby had concluded that I was following up some research on upper-air structures in relation to long-range developments that he and I had discussed in 1939, at the time when I succeeded him as head of the Meteorological Department at MIT. Believing what he had been told, Rossby felt that the British effort in extended-range forecasting was in good hands.

Dr. Stagg's chronology of early actions related to the establishment of a meteorological service for Overlord brings to light two points of considerable interest. In the first place, Sir Nelson's discussion with me concerning a joint forecast center for the Mediterranean theater took place at the time when the Air Ministry was negotiating an agreement with the Americans concerning arrangements for operation Overlord. It seems reasonable, therefore, to assume that the Air Ministry, at that time, was considering parallel arrangements for Overlord and the assault from the Mediterranean theater. Secondly, my visit to North Africa and Italy during January of 1944 coincided with Rossby's visit to London. My connection with the Anzio operation was purely accidental and had no connection with my tasks. There may have been a good reason for my visit to Tunis and Bari, though I never found one. It might not be unreasonable to suppose that my objections to the technical arrangements for the meteorological services for Overlord had been foreseen during the early planning process.

Had I known all of this in February of 1944 when I went to see Riiser-Larsen, my question would most likely not have been hypothetical. Late that night, on my way back to Dunstable, I decided to continue my work in the British service, regardless of what the outcome would be.

CHAPTER 17

Overlord: Planning for Chaos

The tools to him that can handle them.
—Thomas Carlyle (1838)

A few days after my return from Italy, Sir Nelson called a meeting to discuss British arrangements for the Overlord operation. The deputy director, the assistant director in charge of forecasting, Mr. Douglas, and myself were present. Sir Nelson opened by saying, "Now that the *inter-allied* arrangements for Overlord have been agreed upon, it is for us to determine the *domestic* provisions that must be made to enable the Meteorological Office to play its full part in the operation." The words "inter-allied" and "domestic" were "spoken in italics." I took it that this reminder was addressed more to me than to the other participants. The boundary of the area of permissible discussion was abundantly clear.

The assistant director, who was normally concerned with little but housekeeping problems at Dunstable, explained a number of trivial matters, including where the additional scrambled telephones could be located to enable Douglas and myself to take part in Stagg's conferences without being disturbed by other activities. He thought that a corner in the so-called library room could be cleared for that purpose.

No excitement was aroused until the deputy director brought up the question of the time range of the forecasts that Stagg expected to receive from Dunstable. The Americans, we understood, were prepared to provide a routine service with analog forecasts for four to five days in advance; what could Dunstable do? It soon became clear to me, and perhaps also to the others, that Douglas was opposed to any arrangement that would tie him and his branch to any routine issuance of extended forecasts; he maintained that, except on rare occasions, forecasts for more than one day were of little value.

There was much history and long experience behind Douglas's attitude, and his rather rigorous stand must be seen against this background. Traditionally, the Meteorological Office had issued forecasts for one day at a time and then added something called "further outlook." The introduction of these outlooks had not resulted from research on the evolutionary aspects of weather and motion systems, statistical behavior patterns, persistence tendencies, or predictability in a general sense. The outlooks were meant to indicate the forecaster's educated estimate of the broad aspects of the weather sequence for two or three more days. There were no established criteria or guidelines to aid the forecaster in arriving at such estimates. Gradually the outlooks developed into a routine for which many forecasters had but little sympathy. All too often the statements tended to be vacuous, or so generalized that they served no useful purpose. Douglas, understandably, was reluctant to agree to the reintroduction of a service that had failed to produce satisfactory results.

In some respects my views differed from those of my learned colleagues. Less than four weeks ago, in connection with the Anzio operation, I had not only produced four-day predictions to meet a fairly complex set of military specifications, but also expressed a high degree of confidence in the forecasts—and everything had gone well. During my Tromsø years [chapter 3], while forecasting for arctic expeditions, I had given occasional advice for periods of two to four days. More recently, during my Bergen years [chapter 4], I had developed a regular service, with four-day outlooks, for Norwegian vessels operating on the North Sea fishing banks. It seemed clear to me that we could, and should, do something to meet the Overlord requirements.

I could easily agree with Douglas that routine issuance of extended forecasts is of but little value. But a nonroutine service is something else. What one can do when external conditions are favorable need not be reproducible on a routine basis. Moreover, the success depends, to some extent, on the magnitude of the effort that goes into the work and the enthusiasm of the men involved. I felt we had to try; though it might have been wise to play it safe, it seemed far more satisfying to try to be useful.

The scheme that I had developed to provide service to Norwegian fishing vessels was essentially a nonroutine scheme. Three times a week, at fixed times, a message was broadcast to the vessels in harbor or at sea. The broadcasts would give either a statement of the predominant weather that was expected for the coming four-day period, or a statement to the effect that the indications were too indefinite to permit issuance of an extended forecast. At that time, since observations from the upper atmosphere were not available, the criteria used in the framing of these extended forecasts had been derived from a statistical study of five-day charts of the pressure distribution at sea level only. The "no-forecast" messages were issued when the criteria were not well met or when none seemed applicable. After a few years of service, a summary of results showed that in thirty-two percent of the cases "no-forecast" messages had been broadcast, while in the remaining cases definite forecasts, with a high degree of reliability, had been issued. In practical terms, this meant that in roughly two-thirds of the cases the skippers could venture out without any appreciable risk of failure, while in one-third of the cases they would have to use their own judgment and, if they ventured out, they had to face a blind risk.

In the final analysis, the main purpose of weather forecasting is to enable the recipient to minimize risks. A useful step toward this end is to try to isolate the blind risks. Had I resorted to a routine, with issuance of forecasts on all occasions, the value of the service might well have been negligible. I maintained then (in the thirties), and I still maintain, that the would-be long-range forecasters have allowed themselves to stumble into pitfalls. The first of these is the unfounded assumption that predictability exists in all cases; the second is to consider a long-range forecast as a continuous succession of detailed statements. Obviously, a long-range forecast can, at best, only cope with *predominant* conditions and their changes within the forecast period; there is no basis in theory or observation for assuming or asserting that a fluid system like our atmosphere can behave otherwise.

After this necessary diversion we must return to Sir Nelson's meeting. Believing that something could be done to meet the Overlord requirements and to lessen Douglas's opposition, I informed the meeting of the very simple scheme that I had used while serving as regional director for western Norway. The pertinent criteria had been developed for the North Sea and the adjoining coast of Norway. Although there was no proof that they would be applicable to the British Isles, some extrapolation might be possible.

Douglas, who could not reasonably be expected to accept anything that had not been fully described and tested for applicability in the British Isles, remained unenthusiastic. The other participants, looking at things from a management point of view, felt otherwise. They eagerly offered to provide the personnel needed to produce five-day mean pressure charts. In this manner, and in this manner only, did I become involved in the British effort in forecasting for extended periods. This happened about three weeks after Stagg had briefed Rossby and identified me with the responsibility for the British effort in long-range forecasting research.

Sir Nelson's meeting did not result in the formulation of any firm program; perhaps it was not intended for that purpose. Things remained rather vague. I think it was understood, however, that Dunstable would do as much as possible to extend the range. Stagg is probably right in writing that this was done reluctantly and mainly to counter the effects that the American analog forecast might have on the decision making.

At this point, it is necessary to make another diversion, this time into the internal structure of the Dunstable complex. Up to the end of 1941 all forecasting activities at headquarters were concentrated in the branch headed by Douglas; its official designation was M.O.2(a) and it was commonly referred to as the Forecast Branch. Upon my arrival, in late 1941, an already existing embryonic upper-air unit began to expand and was soon "promoted" to an Upper-Air Branch, with M.O.2(c) as its official designation. With this arrangement the forecasting activities became divided between two branches of equal standing. It became customary to speak of "surface forecasting" and "upper-air forecasting." For example, in the daily routine of forecasting winds aloft for bombing and other air operations, M.O.2(a) would provide a map showing their forecast of the winds at sea level. M.O.2(c) would provide maps of the *increase* in wind from sea level to the several upper levels of interest. By adding the two forecasts the winds at any upper level were obtained. This addition, by the way, was considered to be an upper-air responsibility.

There is, of course, nothing in the laws of physics that could justify this kind of arrangement, which, it must be admitted, seemed fraught with undesirable potentialities. Though the forecasting problem is indivisible, the requirements of science and management are not always tuned for resonance; in many cases, two teams are no better than one. In practice, however, our "system" worked quite well, mainly because Douglas and I worked well together, and our separate staffs responded.

There were also deeper reasons for the division of the forecasting work: upper-air analysis and forecasting was a new endeavor in Great Britain. Few meteorologists outside M.O.2(c) had any experience in this field; conversely, few members of my staff had experience in surface forecasting. By the time work on the Overlord operation commenced, I had a little over twenty years of experience in forecasting; Douglas had even more. Both of us had very able and cooperative helpers. A major clue to our success may be found in the fact that though the two branches worked hand in hand, the one branch never interfered in the business of the other.

Although the policy of noninterference was an overriding consideration, a minor aberration became necessary in connection with the Overlord operation. In spring 1944 we began to prepare five-day mean pressure charts in an endeavor to experiment with extended forecasts of the type referred to above. Douglas showed but little interest in the work. Reasonably enough, he could not be expected to add to his load of duties, which had already reached the saturation point. Although no personal jealousy could possibly develop between Douglas and myself, it seemed anomalous that I should be doing "surface analyses." I continued to analyze mean pressure charts, but the conclusions I could draw from them were never discussed with Douglas and his staff, nor was any reference to them made in any of the telephone conferences.

I had two good reasons for adopting this attitude. In the first place, the conclusions that I could draw from the five-day mean charts were often more clearly reflected in the upper-air analyses. Secondly, as we shall see later, the discussions in the telephone conferences became rather unwieldy and it would only add to the confusion if I were to refer to analyses of which the conference participants had no knowledge.

Sir Nelson's meeting brought no "tidings of comfort and joy" to Douglas and myself. I felt that we had been called to hear management decisions rather than to advise on the planning of a highly technical operation. Nevertheless, the meeting proved useful in that several of the complexities of the meteorological part of Overlord began to emerge. From information now available it seems possible to piece together a diagram (Fig. 17.1) showing the meteorological organization that actually existed after the end of the upheavals—at the Supreme Headquarters and at Widewing—to which reference was made in the previous chapter.

Weather forecasts were to be presented over scrambled telephone lines by the three forecast centers (Admiralty, Widewing, and Dunstable). The presentations were to be accompanied by discussions, chaired by Dr. Stagg, with a view to eliminating conflicting views and enabling him to formulate the advice to be presented to the supreme commander and his inner entourage. The commanders in chief of the air and navy

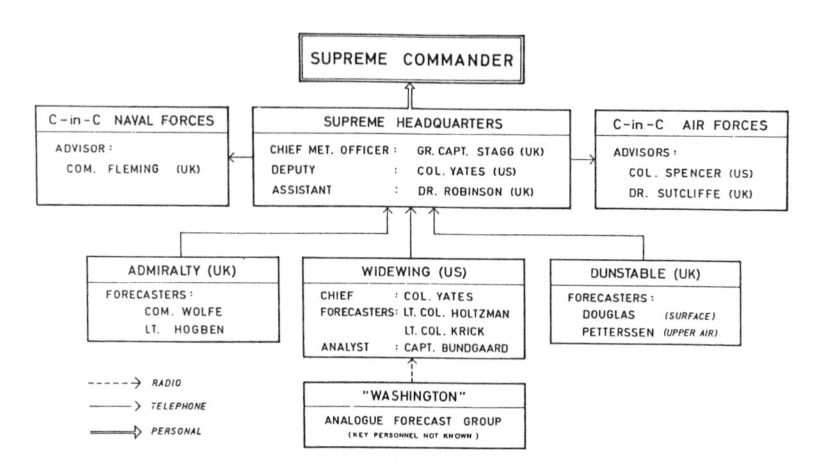

Fig. 17.1: Schematic outline of the meteorological "organization" that was to generate forecasts for the Overlord operation. Admiralty was situated in the center of London, Widewing was situated 20 km farther west, and Dunstable was situated approximately 55 km north of London.

assault forces had their own meteorological advisers; they were connected by scrambled telephone lines to the conference and were generally helpful in asking and answering questions for clarification.

As soon as the system began to function, Stagg, following democratic practices, arranged for rotation of the order in which the three centers were to appear in the presentations. This practice was further extended when it was arranged that the forecasters at each center were to rotate as first speaker for the center to which they belonged.

This latter arrangement caused some inconvenience at Dunstable. Throughout the war, and especially at the Bomber Command conferences, Douglas and I had each spoken as a representative of the branch we headed. In practice this meant that Douglas, without being restricted in any way, was concerned mainly with clouds, visibility, and weather in all their varied aspects, while I concentrated on atmospheric structures, winds aloft, storm developments, large-scale changes, etc. Though the division of responsibility was never strict, a comfortable sort of squatter's rights arrangement had developed between us. However, when the rotation was instituted and I had to appear as first speaker, I found it difficult at times to adhere to our policy of noninterference. Though time was always at a premium, I managed, on all occasions, to obtain at least a hasty briefing on clouds, visibility, weather, and surface winds, either directly from Douglas or from the senior forecaster on duty in his branch.

While the naval component of the organization was a model of simplicity, Widewing seemed rather complex, even as compared with Dunstable. In the first place, Colonel Yates, in addition to being in charge of Widewing, was Stagg's deputy. Since Stagg had no staff and facilities for producing any kind of meteorological analyses, he became dependent upon Widewing for such charts and services as he thought necessary to conduct his conferences.

Other conditions added to the uncertainties. I suspected (but did not know for certain until Stagg's book appeared) that the analog forecasts were not prepared at Widewing: They came from some air force organization in Washington and were trimmed at Widewing for presentation to the conference. According to Stagg,

> Col. Yates's team at Widewing had at least three sources of longer-range prognostic advice. They could use techniques of extrapolation from the recent past into the future, similar to those employed for 12-36 hour forecasts but extended further ahead. They had the forecasts, which came regularly to them from Washington. These were fashioned to take specific account of weather in England and the Channel area, and the methods used to produce them could apparently be adapted to deliver prognostications of any desired degree of detail for almost any period in the future, and, if need be, in any part of the world.

Though, undoubtedly, Krick's claims on behalf of his own analog system were somewhat exaggerated, it seems unlikely that he, in conversations with competent forecasters, ever went to such extremes as those indicated by Stagg. It seems even more unlikely that such exaggerations were endorsed by Colonel Yates and other Widewing meteorologists. Stagg provides some interesting information on the third kind of forecasts that went into the fashioning process at Widewing. Thus, on page 29, we read:

> Widewing also had a unit deployed under Capt. Bundgaard who had a novel and ingenious system which, for all I knew, could have given useful results as frequently as either of the other two lines of guidance available to the U.S. team.

Bundgaard, a very able young scientist, had called on me late in 1943 to seek advice on some research he was then pursuing. His approach had been inspired by certain works of three giants in statistical theory, Herman Wold (Sweden), A. N. Kolmogorov (USSR), and Norbert Wiener (United States). His aim was to derive long-term relationships between upper-air structures and developments at sea level. The idea seemed attractive and Bundgaard left with much encouragement. I did not know, in the spring of 1944, that this half-finished work, which had not been tested for applicability anywhere, was used to contribute to the forecasting for D-Day. In later years Bundgaard's work does not seem to have yielded to development beyond its capabilities at the time it was first put to use.

With reference to the manner in which Widewing contributed to the conferences, Stagg (p. 30) writes:

> The Widewing representatives never divulged at the weather conferences which system or combination they adopted on any occasion, but later experience of their contributions left little doubt in my mind that reliance was placed on the extrapolation and Bundgaard techniques only insofar as they supported the forecasts produced in Washington or on the spot by the Krick procedure.

It is indeed difficult to understand how the scientific merits of different methods could have been usefully described and discussed over the telephone lines in connection with specific cases, unless the methods themselves had been properly explained and tested, and the results made known to the participants. However, if Stagg considered such exchanges over the telephones relevant, it is surprising that he, as the conference chairman, did not make the arrangements that he considered necessary.

Chapter 18

Overlord: Approach to D-Day

Then let us ever bear
The blessed end in view,
And join in mutual care
To fight our passage through.
—Charles Wesley, *Hymns*

To appreciate the magnitude of the meteorological part of the Overlord problem, it will be useful to make brief excursions into the military political landscapes. According to Field Marshal Montgomery,[78] the overriding consideration was the magnitude and time of occurrence of tides in relation to what is called civil twilight. The tidal waves enter the English Channel from the west. As the cross section of the Channel narrows, the wave amplitude increases, with the result that a broad strip of the Normandy Beaches becomes denuded at low tide. The Nazis had built underwater obstacles as far out as extreme ebbs would permit. To stay clear of them it was necessary for the invading force to arrive at maximum low tide.

To achieve surprise, the tide had to occur at dawn, which in early June is about 5:15 A.M. and varies insignificantly from one day to the next. The time of low tide, however, varies by some thirty-odd minutes during the daily cycle. About half an hour was needed by the navy to clear the obstacles. During this interval the landing force would not be in a good position to defend itself. The Allies had enormous air superiority, but in this kind of maneuver air cover would be of little use until about ten minutes after the beginning of civil twilight. In other words, what was needed was some twilight but not too much of it.

A second, and important, requirement was for the navy to be able to land another wave of supplies before dark, also with all obstacles bare. On these and related military considerations alone, the invasion could be attempted only from late spring to some time after midsummer. Two windows of opportunity—each of three or four days' duration—would present themselves in each lunar cycle, making a total of six or seven in a calendar year. The meteorological problem, therefore, was not to choose an ideal weather situation but to decide, some days in advance, whether the conditions would be above minimum requirements on the occasions when the combination of tides and twilight was close to ideal. For purely technical reasons, including military readiness, June rather than May had been chosen, with July as a standby. However, the exclusion of May was not known to the forecasters involved.

The almanac showed that a favorable combination of tides and twilight would occur between June 4 and June 7. On the first of these days, the duration of useful twilight was only fifteen minutes, which was far below requirements. June 5, with about fifty minutes, was close to ideal. June 6 was quite satisfactory. June 7 had about two hours, and though this was too much, it could be accepted. If these three days (June 5–7) were missed, either because of bad weather or rough sea and surf, or because of misleading forecasts, the next opportunity, with suitable tides and twilight (but unknown weather), would come fourteen days later. Although two weeks is a short span of time in a war that was almost 240 weeks old, the horror of having to disembark the troops after they had been briefed and start over again was almost intolerable. According to Montgomery, even this had been included in the plans.

Political considerations were no less weighty. Soviet leaders had long been pressing for an Allied attack from the west, and Generalissimo Stalin had used strong words to describe what he considered to be inaction on the part of his western allies. The Bolero plan, to which reference was made in chapter 12, had held out hope of relief, but a whole year had passed without the hoped-for attack.

Public opinion, in and out of Great Britain, was tense. People suffering under the Nazi occupation were getting tired of bombing and destruction without liberation. The psychological reactions on the Continent and the political consequences everywhere were unpleasant to contemplate, to say the least. Obviously, victory would be ours eventually, but the liberation might come too late. It is against this politico–military background that the importance of the weather services to Overlord must be judged, and it is with these risks in mind that the meteorological organization ought to have been planned.

Even before the military planning began to gather momentum, meteorologists on both sides of the Atlantic tried to determine, from statistical analyses of past data, the time of the year when the conditions would be most suitable for an assault across the Channel. Since the teams involved used assumed military specifications and different sets of data, the results were found to be either conflicting or subject to doubt. As usual, it was easier to demand uniformity than quality.

As the supreme commander's headquarters became established and firm planning got under way, each of the major planning sections produced its own list of military specifications, with corresponding meteorological requirements. In 1942, while working in the U.S. War Department and assisting in the planning of the Plough project (to which reference was made in chapter 12), I obtained the impression that the military planners were idealists in the sense that their emphasis was on the *best* rather than on the *good*. Hardly ever did one hear of that which, should the need arise, would be *barely tolerable*. Once in a while I used to remind my colleagues of Carlyle's words of wisdom: "Let no one measure by a scale of perfection the meager product of reality." They found the quotation, as well as my friendly quips about their Santa Claus lists, both amusing and informative.

Because of the enormous complexity of the Overlord operation, the lists of requirements became both long and numerous. Using these lists, Stagg found that more than a century would have to elapse before the supreme commander (who was then fifty-four years old) could expect an occasion when all requirements would be met. Finding the planners unwilling to revise their assumptions, Stagg set out to compile his own list of realistic requirements. He found that the odds against the occurrence of weather favorable to Overlord in early summer might be as low as twenty-five to one and were unlikely to exceed sixty to one. Though, undoubtedly, this was a step in the right direction, the improvement seemed rather small. According to Stagg, the professional planners showed no interest in his specifications.

Most surprising, perhaps, were the results of a comparison of the individual months then under consideration. Stagg found that May was twice, and July four times, as bad as June. Similar values were found by an American team.[79] Since the weather does not change abruptly on May 31 and June 30, the results, if there is any truth in them at all, could only mean that the unfavorable odds would normally have a distinct minimum somewhere near the middle of June, rise sharply toward May, and rise even more sharply toward July. If this were so, it would not seem unreasonable to suspect that the minimum would be found closer to June 1 than to July 1, perhaps somewhere near the favorable period of tide and twilight (June 5–7) that could determine the selection of D-Day. Since, even in retrospect, the comparison between the calendar months has been considered meaningful, it is surprising that no attempt was made to locate the position of the possible minimum with greater precision. Had such a simple search been made one would have been able either to document the existence and location of the suggested minimum or to shed light on the quality of the statistical analysis as a whole. However, there is no indication that the thinking progressed to include such tests.

As far as the atmosphere is concerned, the "products of reality" are complex rather than meager, and statistical studies may not be taken to represent realistic estimates. In the first place, many of the specifications were hypothetical, and all tended toward idealizations. More important, however, may be distortions resulting from the procedures used in the selection and handling of the data. The atmospheric variables are functionally related through the laws of physics. To determine the occurrences of such variables is really what the science of meteorology is all about. Even slight variations in some of the more important specifications may cause great changes in the statistics on the occurrence of complex combinations. Nevertheless, studies of complex sets will often serve a useful purpose if interpreted on the basis of experience and common sense. One must agree with the great Laplace, who said that statistics is just common sense reduced to mathematics. In the case of the Overlord studies, the common sense element seems to have been rather meager. Even so, there is nothing to indicate that the studies have led or misled anyone, except, perhaps, those directly involved in their making.

Whatever the merits or demerits of the statistical studies may have been, certain facts related to the Overlord operation are indisputable:

(a) May turned out to be incomparably better than June, which, in turn, was somewhat worse than July.
(b) One has to go back to beyond the turn of the century to find a June as bad as that of 1944.
(c) The operation was successfully launched under marginal conditions during a relatively short break in a stormy spell, the actual crossing of the Channel being postponed by twenty-four hours on account of inclement weather, rough sea, and high surf.
(d) From a military as well as a meteorological standpoint, the crucial decision was the postponement of D-Day from June 5 to June 6.
(e) The weather, sea, and surf that caused the postponement to be made, as well as the break that made the launching possible on June 6, were predicted sufficiently early to enable the supreme

> commander to make his decision and issue orders that made it possible (though only barely so) to make full use of the break.

Though the statistical studies provided no guidance to the forecasters, they probably served a purpose in convincing members of the supreme commander's staff that the problem of selection of D-Day was receiving meteorological attention. Since conditions related to tides and twilight were overriding and the number of favorable combinations rather small, the problem became one of forecasting what the atmosphere would do on a few pre-selected occasions.

Ideally, the forecasts would have to cover time spans sufficiently long for the force to be moved, safely landed, and reserves and supplies built up to such an extent that the force could not be thrown back into the sea. If the military commanders were to sit back and wait until all ideal requirements were met, the ideal of *peace forever* would probably be achieved. In warfare risks have always to be accepted. It seemed clear to me that the crucial requirement was to be able to move and land the force safely. From what little information was available, I judged that forecasts covering periods of thirty-six to seventy-two hours would suffice. In my opinion, this was not beyond our capabilities. Fortunately, even shorter forecasts turned out to be useful.

Since so much has been said and written about the weather aspects of the work, it seems appropriate to mention that there was also a marine problem. The wind forecasts had to be translated into predictions of swell from distant sources, the state of the sea along the British coasts and in the Channel, and surf on the Normandy beaches. The marine part of the work was handled with skill and convincing professionalism by the Admiralty center; only very rarely did Widewing and Dunstable contribute to the discussions. At least one part of the problem remained noncontroversial.

Experimental forecasting involving telephone conferences between the Admiralty, Widewing, and Dunstable began in late February, only a few days after the arrival in England of key personnel of the Widewing team. Though it seemed clear that our American colleagues were lacking in experience of the weather in western Europe, the analog system seemed to work fairly well, at least up to forty-eight hours or so, as long as typical winter storms dominated the Atlantic region. Moreover, some of the junior forecasters at Widewing had served in England for many months. Undoubtedly, their experience, though not extensive, must have proved useful on many occasions.

During the initial experimental period, telephone conferences were relatively infrequent and the discussions rather relaxed, except, perhaps, as regards forecasts for three or more days. All of us seemed to work on the assumption that D-Day was still far off and little or no sense of urgency prevailed. Someone, somewhere within the system, needed to gain experience and we were there to simulate something that we would have to face at some later time.

Normally Douglas spoke for Dunstable, with emphasis on visibility, clouds, and weather for a day or two, while I supplemented with information on the grand currents aloft, upper-air structures, and related aspects, much as we had done at the Bomber Command conferences throughout the war.

While it was relatively easy for me to keep out of controversial matters (if I so desired), I was disappointed to find a general lack of interest in the evolutionary aspects of upper-air structures, interactions between the upper and lower regions, and their relation to the large-scale aspects of storm development. Some general guidance could be obtained from the pioneering work of J. Bjerknes and Palmén (published since 1930), Rossby's theory of long waves (1939, 1940), and the results of the statistical investigations of upper-air conditions, as related to sea-level patterns, conducted at MIT since about 1935. I had been fortunate enough to be closely associated with Bjerknes, Rossby, and the MIT group, and I felt that I was well informed on German contributions in this field up to June 1939. True, none of these works had been developed into a set of simple arithmetical rules; nevertheless, they added up to a fair volume of the mental equipment of those of us who looked upon forecasting as a problem in three dimensions.

Lieutenant Hogben, always keen and enterprising, had spent a week at Dunstable to become familiar with the work we were doing in the Upper-Air Branch. Even this very short visit proved useful in establishing some kind of common language with the Admiralty team. Dr. R. C. Sutcliffe, who, with much eloquence and insight, had contributed so valuably to the Bomber Command conferences, had now been posted to the staff of the air commander in chief. In this capacity he had little but skeletal data and analyses to look at. Instead of being among the main contributors to the D-Day conferences, he had now been relegated to a listener; only rarely did he contribute to our discussions.

I had long assumed that D-Day would be chosen in May, which, as most experienced forecasters would agree, has many unique advantages. Normally, May has a high frequency of deep high pressure systems (or anticyclones) sitting over the British Isles or somewhat farther to the west. These anticyclones, which reach up to great heights, are very stable and almost stationary. Meteorologists refer to them as "blocking highs" because they block the eastward passage of Atlantic storms. Should such a high settle down over or near the British Isles, our problem of forecasting for D-Day would be enormously reduced. Uncertainties in analyses of the conditions over the North Atlantic (where observations were extremely sparse) would be of little importance and a source of controversy between the different centers would be eliminated. Sea, swell, and surf would cause no worry; general forecasts for a few days in advance would not be much of a problem; and the precise choice of D-Day would depend mainly on a forecast of clouds and visibility at low levels. Surely Douglas's skill in this area would suffice. Events, however, took a different course.

At the beginning of May there was a sudden increase in the frequency, length, and urgency of the conferences. Since I did not know that May had been ruled out, I began to focus attention on the first of the favorable tidal periods of May. Though my lodgings were only five minutes walk from the station, I had had a bunk installed in my office. Since Douglas and I had heavy duties apart from Overlord, it seemed wise to be prepared for a long period of irregular work. With this added load, we lived, as it were, from hand to mouth, worked and worried, and became increasingly concerned about the trends of the conference discussions. From Stagg's book we learn that some of this increase in conference activity was introduced by him to bewilder anyone who might try to predict the choice of D-Day. Apparently it was considered inadvisable to extend confidence to Douglas and myself. In the meantime, both of the favorable tidal periods of May passed and D-Day remained elusive.

The increase in conference activity served at least one useful purpose: The weaknesses of the organizational framework became abundantly clear, even to the extent that failure could not be ruled out as a possibility. In the first place, owing to the scarcity of reports, especially from the North Atlantic, the analyses at the three centers differed significantly, particularly in cases of incipient or potential storm development and unusual upper-air structures. Though such differences remained at the root of the problem, the telephone discussions were invariably centered on the forecasts themselves, with only scant attention to the bases from which they were derived.

Secondly, substantial misunderstandings arose from the fact that there had been no exchange of pertinent information on the terminology and methods used by the different teams. This, together with the uncertainties in the analyses, tended to turn the conferences into statements of opinions rather than informative exchanges of deductions logically arrived at from some accepted basis.

A major and persistent difficulty arose from the circumstance that the chairman had not the data, facilities, and staff, nor the scientific block and tackle, that could enable him to make forecasts of his own or to pass real judgment on the quality of the advice he received. In the early phases he had to rely on what he heard over the telephone lines. Later, he was able to obtain copies of the analyses produced by Widewing. Though he showed but little confidence in their methods, he became completely dependent on their charts as guiding material for his chairing of the conferences.

The difficulties were especially bothersome when it came to discussions of forecasts beyond thirty-six hours. Widewing, using the analog technique, was always on the generous side. Once they, or the analog group in Washington, had identified what they considered to be the "best analog," it was thought that the weather for the ensuing days would repeat the sequence that followed the analog; therefore, they had no difficulty in pinpointing the details for several days in advance. To Widewing, a long-range forecast was nothing but a continuous succession of statements of details, or, one might say, a succession of short-range forecasts. Since this was the type of forecast that the military desired, the chairman, being under pressure and having nothing but Widewing material for his guidance, kept pressing for comments on "the Channel weather" for each day of the five-day period.

The Admiralty had made a cautious and entirely sound study of certain weather types which must have been useful as background information and for indoctrination of new personnel, though, as things developed, its influence on the discussions was probably small. Douglas, as we have seen, was opposed to the issuance of forecasts for periods in excess of thirty-six to forty-eight hours. But being very cooperative and anxious to preserve the spirit of the conference, he commented generously on the Widewing extensions; sometimes he compromised beyond his better judgment and regretted his compromising when the conference had come to an end.

To me, a long-range forecast meant a statement of the *predominant* weather for some period after the time to which the short-range forecast referred. I was unwilling to set a standard time limit to such forecasts; whether it should be thirty-six, forty-eight, or *x* hours would depend on the meteorological situation itself. In no circumstance could I agree that useful predictability existed routinely. However, such questions were never discussed; though we all spoke the same language (I with what is often called a Minnesota accent), the meaning of important words and concepts remained obscure.

Some time in May, Douglas showed me his notes on how the Dunstable forecasts had fared after conference discussions and adjustments through compromises. It was clear that the compromise versions were less accurate. At this time we knew that Stagg's assistant had checked on the extended forecasts and found but little evidence of skill beyond forty-eight hours. It was clear that, on the whole, the stretching of the timescale of the forecasts and the compromising had served no useful purpose.

So far, I had considered the Meteorological Office involved more through Douglas than myself. Also, I had developed the impression that the chairman did not like to hear much about upper-air analyses. His interests seemed to be centered on the details of "Channel weather" for each day of the forecast period. Clearly, the conference procedures had led to divergence, rather than convergence, among the three forecast centers. Some changes seemed in order. I suggested to Douglas that we take a firmer stand and not compromise as readily as we had in the past. Both of us had long experience in forecasting for western Europe, while most of our able colleagues had only recently come on the scene. Though responsibility cannot be apportioned according to any standard rule, it seemed clear to me that Dunstable ought to recognize more than an even share. What we could do immediately was to bring more firmness into the Dunstable presentations and, thus, leave the chairman less room for compromise interpretations of our contributions. As for myself, I decided to take a more active part in the conferences. The only way I could do so was to expand my references to the upper-air analyses.

No account of the forecasting for D-Day will be reasonably complete without a description of Douglas and the manner in which his mind worked. In his book, Stagg, who obviously admired Douglas and relied heavily on his judgment, says (p. 54), "Douglas was apt to leave an impression of uncertainty, hesitancy and lack of confidence. . . ." His presentations were characterized by "many fine shades of probability conveyed in a rambling, sometimes seemingly inconsequent, monotone." Undoubtedly there is much truth in Stagg's description, but more needs to be said to balance the picture. Douglas was a genius, though not a well-organized one. Eloquence was not one of his outstanding talents. Unlike others in his class, he was completely lacking in aggressiveness, and generally short of self-confidence. At the age of ten he had read a popular book about weather, and under the influence of a nature-loving teacher, he began to keep a weather diary. In his sixteenth year he made his will felt and was allowed to transfer from classical studies to science. He then became a regular subscriber to the daily weather charts published by the Meteorological Office. In the First World War he chose to serve as a weather observing pilot and became increasingly fascinated by what he saw: the weather phenomena in all their complexities.

Douglas never simplified; he was not interested in blueprints, models, and abstract theorems. His domain was the entire complex phenomenon, seen as an entity. In understanding the totality of interwoven processes, Douglas had gone farther than any forecaster I have ever known. Douglas was a Solomon and not a Hammurabi. His wealth consisted more of wisdom than of science. Neither meteorological journals nor analects can record what he knew.

There is one episode, not a very important one, that illustrates well how his mind worked. On a certain Tuesday we were rather late for one of the innumerable weather conferences. I was to be the first speaker, and we had not had time to discuss the weather situation. As we rushed to the telephone I asked Douglas, "Have you thunderstorms in the picture?" A mumbled "yes" reassured me. When I asked "When?" he said "Thursday." I was pleased and surprised. I had been following a mass of cold air aloft, projecting southward into the subtropics and moving slowly eastward across the Atlantic. It was now near the Azores meridian, and in two days it could be expected to be over England. With cold air aloft and daytime heating over land (and for other reasons) thunder could be expected. But Douglas, being busy with his own work, had not seen the upper-air data and charts I had used. When the conference was over I asked him how he had come to his forecast of thunderstorms. His answer was a typical one. On a certain day (I believe he said it was in 1911), a similar case had occurred, "but then the outbreaks were in Brittany; this time I think it will be farther north." I broke off there, for had I continued, he would probably have hacked our forecast to pieces and mentioned many other things that could also happen. With Douglas, the discussion often started the moment agreement had been expressed.

Thursday morning arrived with perfectly clear skies and no indication of what we had predicted. During the forenoon I often went out into the field beyond the overhead camouflage to see if thunder clouds had appeared on the western horizon. At about 11:00 Douglas and I happened to meet in the open. Neither of us was happy; soon there would be another conference. The tension within me was relieved when I heard Douglas saying, "If we only could heat it up a bit." Well, we couldn't. But shortly before 2:00 P.M., thunderheads began to appear and soon storms broke out over much of England. But Douglas was no longer interested; there were always new problems. In arriving at his forecasts Douglas often moved like a sleepwalker: He found what he was looking for but his path was often peculiar and tortuous. He often distrusted what he had found.

Toward the end of May the general weather situation over much of the Northern Hemisphere began to change. Although the "Channel weather" did not respond immediately, the signs were ominous. It so happened that the consequences of the change became critical on the day planned for the landing. At the root of the change was the production of enormous amounts of deep arctic air, especially on the Atlantic–American side of the Pole. Normally, as the arctic air finds its way down to subtropical latitudes, it absorbs so much heat on the way that it arrives there as a welcome whiff of cool, refreshing air. But now, as Overlord was approaching, the southward passages were blocked, and hardly any arctic air streamed across a line extending from the Canadian border eastward across Labrador, Iceland, and the north coast of Europe. Most of the arctic air that found its way down to subpolar latitudes was returned to the Arctic for further cooling. With so much production and but little export, an enormous high pressure area developed and covered the whole region from the Canadian Rockies, across Greenland and Spitzbergen, and eastward to Novaya Zemlya. How far it extended toward eastern Siberia and Alaska we did not know.

Meteorologists accustomed to sea level charts only saw pressure rises in Iceland and Greenland. They wondered what might have caused them and what they might lead to eventually. Others, being also familiar with upper-air charts, foresaw a vast increase in the north–south temperature difference and a corresponding increase in the westerly winds aloft over much of the hemisphere. Men familiar with the works of the MIT group saw a development toward a "high-index" circulation, meaning strong westerly winds aloft, and fast-moving storms under these winds. Other and related conclusions would come from the works of Bjerknes and Rossby on upper waves and from many German writings on *Steuerung*, meaning the steering of sea level systems by the currents in the upper atmosphere. None of these works and recent additions to our knowledge could tell us much about such details as "Channel weather," but they certainly would provide guidance for assessing the evolution of the large-scale situation and the predominant weather systems for a few days at a time. Also, familiarity with upper-air structures and analyses would help materially in arriving at more meaningful analyses of the current weather situations at sea level. The atmosphere is a unique physical system: All layers are interrelated, and no particular level or layer can be said to be more important than others.

In some respects May 28 became a turning point in our work. The supreme commander, with key members of his staff, was about to move to the advanced command post near Portsmouth, adjacent to the headquarters of Admiral Ramsey, the commander in chief of naval forces. On this brilliant spring Sunday, Stagg and Yates drove down to Portsmouth, leaving their assistant, Dr. Robinson, to "hold the fort" at their London headquarters. Desk space, telephones, and other facilities were provided for them by Commander Fleming in his improvised and already crowded weather station which had been set up to serve the admiral. From then on Stagg and Yates became dependent on the analyses provided by Fleming's unit while their assistant and confidant in London continued to be guided by the Widewing charts. This dispersal, in itself, did not add much to the complexities of the telephone conferences, which had already reached a high level. Upon his arrival at the advanced command post, Stagg lost little time in calling a conference; though the weather situation was discussed in general terms, the main purpose seemed to be to ascertain whether the conference mechanics worked smoothly.

We conferred again late the same evening. Stagg now asked for forecasts for each of the coming five days and further outlooks for three more days, to include D-Day, June 5. Speaking for Dunstable, I summarized the thirty-six-hour forecast prepared in Douglas's group and went on to describe the broad aspects of the developments for the remainder of the five-day period. In my opinion the upper-air situation across the Atlantic was approaching an unstable state and deterioration would set in toward the end of the five-day period (June 1 and 2). Krick, speaking for Widewing, foresaw a different kind of development, with the weather remaining quiet throughout. At the Admiralty, Woolfe joined Dunstable in foreseeing deterioration. Although some of his details seemed somewhat hypothetical, they were certainly within the realm of probability. I found his discussion inspiring and was glad that two of the three centers were in substantial agreement. After the conference I carefully reviewed Krick's forecast and came to the conclusion that, but for the upper-air structures and my projected increase in the upper-level westerlies right across the Atlantic, his forecast did not seem unreasonable. However, the upper-level winds were there, and they were strengthening.

Judging from Stagg's account the absence of agreement between Krick and myself, so close to D-Day, was disturbing in a very high degree. Thus, on page 64, he writes:

> Should I ask the two weather services, U.S. and British, to arrange that for the remaining all-important series of discussions Krick and Petterssen never participate at the same sitting? This, I realized, was impracticable . . .

Though opinions may vary as to the practicability of such a request, the point of interest now is that no action was taken (except to record the thought for its historic value). So I was allowed to sit with the other forecasters, take part in their discussions, and express my views on upper-air structures, storm developments, and related matters in terms of my own choosing.

My notes have no reference to the conferences on Monday May 29, probably because it was not my turn to speak for Dunstable. Having worked through much of the night, it is unlikely that I attended the morning session. Judging from Stagg's account, optimism seems to have prevailed. At the morning conference, the deterioration that had caused the controversy the previous evening had now been reduced to "a risk of minor and temporary disturbances Friday and Saturday (June 2 and 3)." It seems remarkable that such specific phrases as *minor* and, especially *temporary* were used in a forecast for days so far away. Though the word "temporary" could indicate that either better or worse conditions were likely to follow, most readers would probably discern a note of optimism for D-Day.

The evening conference on May 29 seems to have reflected sustained optimism; Stagg (p. 65) reports that the spokesmen for the three centers were agreed "that a serious breakdown at the weekend [June 3 and 4] seemed improbable." Since Douglas, as we have heard, was opposed to forecasts for periods longer than thirty-six to forty-eight hours, it seems reasonable to assume that this forecast for six days in advance reflected absence of opposition rather than positive concurrence by him. The optimistic views had won the day.

On Tuesday May 30 the conference was divided. Arguments for deterioration toward D-Day were again expressed, though, not unreasonably, the timing of the onset was left uncertain. Saturday and/or Sunday were mentioned. Widewing, however, continued to maintain that the Channel area would be protected by a northeastward extension of the semi-permanent high pressure area near the Azores. The "battle of the experts" continued to seesaw, with results that, to some extent, depended on which of the centers led off the discussion and who was the spokesman at each of the centers. Douglas and I probably had certain advantages over the other participants; we had more than age and experience on our side. Although I had been commissioned a lieutenant colonel in the Norwegian Air Force, I used the uniform only for travel purposes. We were both civilians working in the remote Dunstable center. Since there were no officers of exalted rank anywhere near us, it was easier for us than for our colleagues to remain uninfluenced by what was expected or desired. Especially when observations are sparse and one's judgment is taxed to the utmost, it is easy to become influenced by external conditions and pressures. The lessons I had learned in the late twenties, when forecasting for arctic expeditions, had not faded from my memory.

As D-Day approached and the large-scale weather situation worsened, I became increasingly concerned. In the conferences, Stagg continued to press for consensus day-to-day predictions of "Channel weather" for five days in advance. So far, agreement had hardly ever been achieved. In management, decisions are normally reached via one or more of three different avenues: persuasion, integration, and imposition. In our case the first two methods had not proved productive; the third could not possibly succeed unless the chairman were in a position to make forecasts on his own.

At no time did I doubt that the requests for agreed upon predictions for periods covering five days were seriously meant. However, judging from Stagg's account, it appears that the situation may have been more complex. During a brief period in February 1944, between his demotion and reinstatement as chief meteorological officer to the supreme commander, he found time to study a series of sea level weather maps. He soon came to the conclusion that existing forecasting techniques were faulty. On pages 38–39 of his book he asks:

> Should I make a clean breast of my doubts and tell my SHAEF [Supreme Headquarters Allied Expeditionary Force] superiors that there could be no scientific justification for pretending to forecast with any degree of confidence or useful reliability English weather (and more especially Channel weather) for more than two days ahead? Or should I suppress my scientific scruples. . . ?

With no lack of frankness, Stagg explains that after having considered the circumstances related to his demotion and the fact that he was no longer in charge, he decided on the second course. However, a few days later, when he was reinstated in the top job, he had to find a different justification:

> . . . the reason I gave myself was this: as chief of the section I would be in a position to ensure that the advice from the British meteorological side was used to leaven the contribution from Widewing: whereas if I adopted a strictly scientific attitude and handed over my task to Col. Yates the preponderance of forecasting opinion might well pass to the U.S. meteorologists. . . . (p. 39)

In describing how he arrived at some kind of balance between the principles of scientific research and the dictates of management, Stagg refers to what he calls "Douglas's dictum," which he calls on for support of his own findings, namely, that forecasting for intervals longer than forty-eight hours is scientifically unsound. It is remarkable that he arrived at these results only a few days after he, in his briefing of Rossby, had identified me with responsibility for the British effort in long-range forecasting research and explained "what was going on with great urgency at Dunstable" to meet the Overlord requirements. As we entered the final phase—the selection of D-Day—it was, perhaps, fortunate that we were ignorant of the existence of these undercurrents.

CHAPTER 19

Overlord: the Postponement

Nothing is so annoying as
to be hanged in obscurity.
—Voltaire

On Wednesday, May 31, with the tentatively selected D-Day only five days away, the weather to be expected on June 5 began to dominate every conference. At the morning session Dunstable, with Douglas as spokesman, repeated and added to our earlier arguments for deterioration of the large-scale situation over the weekend, but the conference came no closer to agreement than on the previous evening. A session late in the day brought some progress, in that the Admiralty now agreed with Dunstable on bad weather over the weekend of June 3 and 4, continuing through the fateful June 5. At this time the emphasis was on the end rather than on the beginning of the forecast period. This, together with the worsening of the large-scale weather situation, added greatly to our difficulties.

From the very beginning of weather forecasting (about 1855) until about 1965, when versatile electronic computers began to make their influence felt, simple extrapolation of the movement and rate of development of weather systems remained an important technique. However, the usefulness of this method is limited largely to forecasts for short periods, say, up to thirty-six or, at most, forty-eight hours. The reason for this limitation is readily understood. The average life span of a major weather system in middle latitudes is normally of the order of four to eight days, with the result that entirely new systems may form and acquire dominating proportions within a five-day period. Thus, on the basis of extrapolation *alone*, there can be but little justification for trying to predict the day-to-day variations of the weather over periods of five days.

In forecasting for several days in advance, one has to search for evidence of potential and incipient developments. And it so happened that the two major storms that dominated the scene of June 5 did not even exist on May 31. The same is true of the relatively quiet area between these storms that made it possible (and only barely so) for the assault to go ahead on June 6, after a postponement of twenty-four hours.

What did exist on May 31 was an enormous mass of cold air trapped in the Arctic, a deep and strong zone of temperature contrasts in the midlatitude belt, and, correspondingly, a very strong and almost straight westerly current in the upper atmosphere, sweeping all the way from the Great Lakes region to the Baltic Sea. This current had the ability to produce storms and to steer them eastward, with strong winds at sea level in the area of Overlord interest. The *predominant* weather for the coming five-day period, I thought, was clear, but the passage of the individual storms and the breaks between them could not be timed several days in advance. Here, there was a clear need for an understanding of the difference between long-range and short-range forecasts. Much confusion—through all conferences—resulted from the absence of definitions and clarifications of terms.

However, on May 31, Widewing found that the Washington analogs supported their earlier views that an offshoot from the high pressure area centered over the Azores (Fig. 19.1) would develop, extend northeastward toward Ireland, and thus protect the Channel against any direct onslaught of the Atlantic storms.

On the face of it the two views may seem quite incompatible, and yet it was difficult, even for experienced forecasters, to discard one in preference for the other. If Widewing was right, the Atlantic storms would be pressed farther to the north, toward Iceland, and though strong winds and low clouds might spread over parts of Scotland, the "Channel weather" might be tolerably good. The question was really not one of *storms* or *no storms*. In practical Overlord terms, the question was how far to the north the storm centers would go; if they moved sufficiently far to the north, their speed and intensity would be immaterial as far as Overlord was concerned. Had the discussions been focused on this issue, rather than on the day-to-day variations in "Channel weather," some convergence of the estimates might have resulted.

From my upper-air analyses, I found support for the view that the "birthplace" of storms would remain near the Great Lakes region (where the upper current curved from a northerly to an easterly direction) while the "graveyard" would be in the Baltic area (where the upper current fanned out and weakened). Though I had little confidence in the development of a "finger of high pressure" from the Azores, as foreseen by Widewing, I kept this possibility in mind.

Although, in retrospect, it is clear that our Widewing colleagues were mistaken, it seems unfair to say that their views were unreasonable. Defense for their attitude may be found in four different areas. First, a "finger of high" did develop, and reached some prominence on June 2. But under the influence of the strong upper current, it proved unable to maintain a position as far north as the Channel. Instead, the "finger" swung eastward, across Spain and southern France. In this position, it backed up the strong westerly winds that prevented the crossing during the night of June 4–5.

Second, owing to lack of observations from the Atlantic, all three centers experienced difficulties in maintaining consistent analyses of current situations. It seems reasonable, therefore, to assume that the analog selection, particularly by the remote Washington group, must have been somewhat unreliable.

Third, and perhaps more fundamental, was the fact that the large-scale weather situation in early June 1944 was abnormal in the extreme, with the result that the best analog, chosen from a series of daily charts covering a period of forty years (1899–1939), proved not to be good enough. In fact, it was not until the early sixties that a reasonably good analog did occur. In long retrospect, it seems safe to say that, in abnormal cases, no analog system will be of much value unless it has been developed to include similarities in vertical structures as well as in time sequences. The atmosphere has a fine memory of its recent past; if two situations, however widely separated in time, are truly similar in all three dimensions and in recent history, there is no known reason why they should not have similar futures. The last word about analogs has not yet been spoken, nor will it be until a sufficiently long series of upper-air charts has been compiled. A century or more of such charts may well be needed to determine the full use that can be made of analog techniques.

A final argument against a harsh judgment of the Widewing views may be advanced on a comparative basis. Though Dunstable consistently opposed these views, prominent Meteorological Office forecasters serving elsewhere supported Widewing up to the very end, when Overlord had to be postponed. The Widewing forecasters, being short of experience from western Europe, must have derived much encouragement from this support.

On June 1 it was my turn to speak for Dunstable. From now on the telephone conferences became so frequent and so lengthy that it was

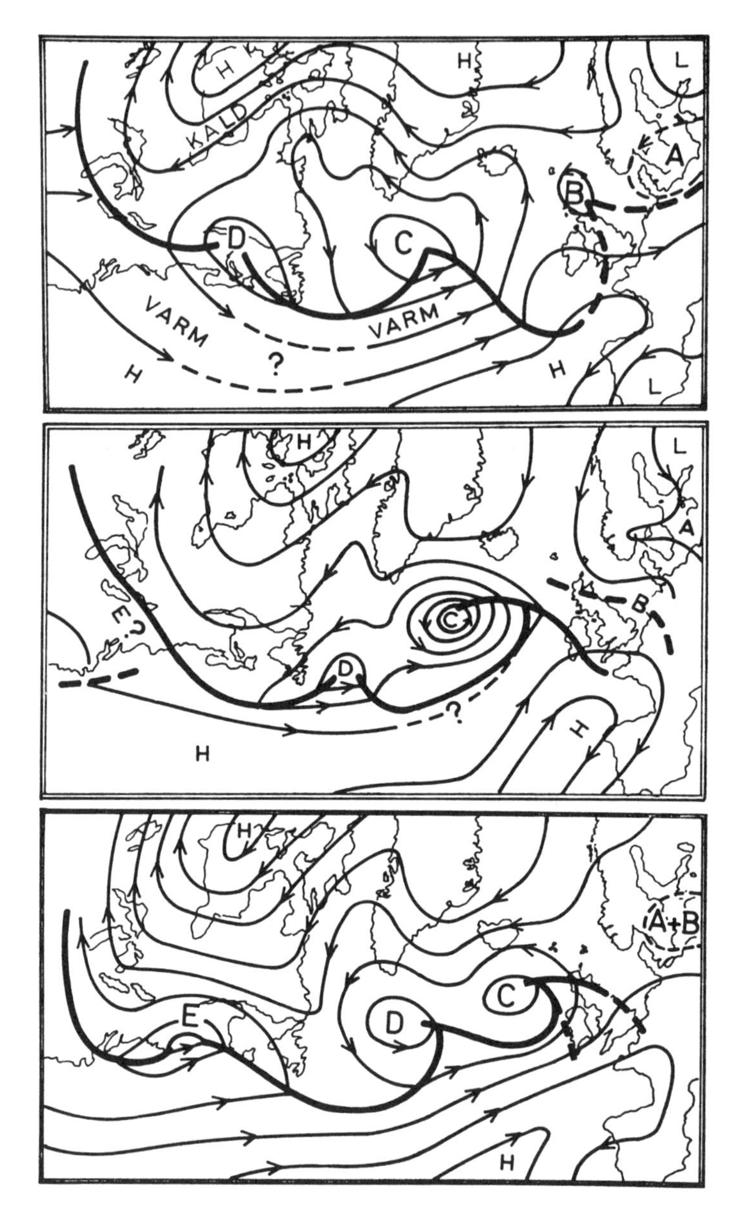

Fig. 19.1: Surface weather maps for June 1, 2, 3, 1944 at 1300 GMT.

The maps depict the main features of the development of the weather during the days prior to D-Day. The air currents close to the surface of the earth are of particular interest. Lines with arrows show the direction of the wind. An inverse relationship exists between the force of the wind and the space between the lines: the narrower the channel, the stronger the hard for Douglas and myself to find time for work and needful rest. Though we could have taken turns at the sessions, we preferred to attend together. Douglas, with his skill at forecasting clouds and weather in England, was simply irreplaceable, and he, probably, felt that my upper-air work was useful.

It was with a feeling of hopefulness that I prepared for the morning conference. The large-scale situation had developed as expected. The temperature contrasts in middle latitudes had increased. So had the upper westerly current, particularly on the American side (Fig. 19.1). Storm D was a new creation, and the temperature contrasts were sharpening in its rear. The Great Lakes region had to be watched for signs of new developments. Surely, I thought, Widewing and I would have no difficulty in agreeing on what was likely to come out of North America.

Speaking for Dunstable, I referred to the good behavior of the upper air and summarized the forecast of sea level conditions as prepared by Douglas and his group. At this time it was difficult for me to see how any new controversy could develop. Widewing, however, thought they had found evidence of a new surge of rising pressure to reinforce the "finger" on its northern side. They looked forward with considerable optimism to

---

wind. The thick line connecting the low pressure centers (A, B, C . . .) is the polar front, that is, the transition zone between polar and tropical masses of air.

The upper map refers to June 1 at 1300 GMT. An area of high pressure is normally in place from Bermuda over the Azores and toward Portugal, but in this case the center is located east of the normal position—a circumstance that favored Widewing's optimistic view of the development.

The oldest members (A and B) of the system of bad weather are now weakened, while the young storms on their way from North America are becoming increasingly intense. The polar front is strongest west of storm C, and one can expect the storm centers over the Atlantic to move quickly eastward due to a very strong wind in the upper layers of the air.

The middle map represents the situation on June 2 at 1300 GMT. Storms A and B are now breaking up; storm C has become stronger; an offshoot of the high-pressure zone above the Azores has now developed (as predicted by Widewing); and a new storm (E) is under development over Pennsylvania.

The bottom map represent the conditions on June 3 at 1300 GMT. Enormous masses of arctic air flow south and enhance the storms above the Atlantic. It was with this map and some later observations as a basis that the meteorologists advised the postponement of D-Day.

June 5. With very few observations from the Atlantic it was difficult to support or refute their findings. In fact, there seemed to be some evidence of a new surge; it was all a matter of degree and, therefore, of discussion and argument. It will be seen from Fig. 19.1 that a surge of high pressure had developed by June 2, but the Atlantic storms were then very powerful. From now on the finger of high pressure would swing southward and give way to the westerly winds from the Atlantic. Though the Admiralty expressed broad agreement with Dunstable, the conference remained divided, and the cleft widened as the day wore on.

In Stagg's account we find several references to the discussions on June 1. Some of them are remarkable in that they clearly reveal the background of the conference scene. With reference to the Dunstable views he found that they, "particularly when presented by Petterssen, seemed to be to be unnecessarily and unjustifiably gloomy."

With some justification I could have been criticized for *not* being sufficiently "gloomy," for the weather and winds during the night of June 4–5 turned out to be even more severe than Douglas and I had predicted. As a result of my "unmitigated gloom" and Widewing's "undiluted optimism," Stagg decided that it would be his duty "to hold a very fine balance among the various opinions until the last moment. . . ."

Having passed judgment on my prognostications Stagg goes on to comment on the quality of my diagnosis. On page 75, we read:

> . . . inter-relations between upper atmosphere and surface weather events, [which] should have followed logically from what we knew of the state of affairs 18,000–20,000 feet above our heads, did not match what was actually in progress at sea level. To my mind important localities of rising and falling barometer on the ground did not occur where the pattern of upper winds and temperatures predicted they should occur. For that and similar other reasons my confidence in Petterssen's diagnosis was not very high.

Apparently Stagg attaches much importance to this verdict, for, on the next page, he repeats it in more general terms:

> Dunstable, on the other hand, and particularly Petterssen, used theory to relate the flow patterns in the upper atmosphere to the changes and developments in the high and low pressure areas at surface level; each case was treated empirically (*sic*) and, in theory at least, Petterssen needed only to forecast the changes likely to take place in the upper atmosphere to be able to define the broad future pattern of the ground level chart.

These statements, which have no basis in science, can only be taken to reveal Stagg's own ideas of how the atmosphere works and forecasts are made. They indicate, perhaps better than anything else, why the conferences so often failed to produce useful results.

Stagg complains (page 55) that his work became especially burdensome when (as he judged) Douglas and I did not agree on future events. When such disagreements occurred at Widewing, the problem was not serious, for "At the least hint of discrepancy between Holzman and Krick at any conference, Col. Yates saw to it that the differences had disappeared by the next meeting." At Dunstable, however, there was no such superior officer to enforce discipline. Being unfamiliar with the division of work and responsibility at Dunstable, Stagg often failed to understand that Douglas and I had arrived at perfectly compatible results via different paths. Stagg often interpreted such differences as being substantive.

Though Douglas and I took turns as first speaker for Dunstable, both of us attended almost every conference and contributed freely to the discussions. Naturally, Douglas and I could only express ourselves in terms that reflected our different fields of responsibility and interest. Apparently, this caused Stagg to become concerned about what he took to be lack of uniformity in the thinking at Dunstable. Thus, on page 77, we read:

> . . . Douglas' long experience and insight served, when it was allowed the chance, to keep Petterssen's theoretical empiricism (*sic*) in check. My doubt was in knowing what Douglas himself thought about matters and I would greatly have valued a private talk with him had not circumstances and protocol ruled that out.

These and other outbursts by Stagg are remarkable, not least because they have no discernible connection with the discussions that took place on June 1. What could be the purpose of this accumulation of vituperations at this critical time? Did Stagg, like the Greek tragedians, deem it useful to engage in goat-singing[80] to draw attention to the drama that was about to unfold and to emphasize the importance of his own role?

Stagg's extraordinary pronouncements would have deserved no comment here had not the circumstances that surrounded our work been so exceptionally grave. On June 1, the greatest military force ever assembled stood poised at the shores and airports of Great Britain waiting for the signal from the supreme commander to launch the assault across the Channel that would begin the liberation of Europe. As the moment of decision drew near, it became clear to all concerned that the success of the entire operation depended in no small measure on a single factor—the weather.

June 2 turned out to be another day of seesawing discussions and indecision. Dunstable, with Douglas as spokesman, foresaw strong winds and much cloud in the Channel. Along somewhat different paths, Douglas and I had found that the Atlantic storm centers would continue to move eastward, toward southern Scandinavia. Widewing, on the other hand, could now, with justification, point to the fact that their "finger of high pressure" had been reinforced on its northern side (Fig. 19.1). Their analogs indicated that this trend would continue. As a result, Storm C, they thought, would be forced northward, toward Iceland, and the Channel would have "operational weather" on June 5. Judging from sea level charts only, and having regard to the scarcity of observations from the Atlantic, both views could be defended. Again, it must be admitted, external advantages were on the Dunstable side. As a matter of routine, having been developed mainly for night bombing operations, charts for seven different levels were analyzed at Dunstable. We were probably in a better position than the other centers to judge what was going on in the atmosphere as a whole.

Stagg, apparently, found it difficult to maintain "a fine balance" between the "undiluted optimism" of Widewing and the "unmitigated gloom" of Dunstable. At the supreme commander's conference he presented what he calls a "hybrid forecast," which, he says, "quite openly leaned more heavily to the Dunstable side but was not as consistently gloomy." He adds, "It was probably a bad forecast." In connection with the presentation of the "hybrid forecast," Stagg reports that his deputy, Colonel Yates, threw in a suggestion that there might be "a complete collapse of the weather situation on Tuesday" (June 6). Stagg goes on to explain that Yates, in saying so, was probably motivated by loyalty to his men at Widewing who had been pressing for June 5 as favorable for the assault.

The confusion did not end there. A preliminary conference in the afternoon of June 2 brought no convergence toward an agreed forecast. However much the Admiralty tried to steer a middle course, Widewing and Dunstable maintained their views, Widewing for and Dunstable against operational conditions during the planned crossing with landing on the Normandy Beaches in the early hours of June 5.

A commanders' conference was scheduled for 9:30 P.M. on June 2; the meteorologists began their discussions at 8:00 P.M. As before, the meteorological conference remained divided. Stagg, however, claims that he, quite independently, foresaw some kind of development, somewhere between Iceland and Hudson Bay (a stretch of 3,000 miles), that might reinforce one of the three Atlantic storms and sweep away the "finger of high pressure" within a matter of a few hours. Stagg did not share his visions and views with the forecasters, for the development foreseen by him "was too simple for any of the experts and certainly conflicted head-on with the views of Widewing's forecasters and those who reasoned on like lines." Stagg now briefed the supreme commander and, though he does not reveal the wording of his summary forecast, we are told (p. 88) that it ". . . was more pessimistic—much more pessimistic—than all but *perhaps* (my italics) the Dunstable participants would have had me present; and Yates, I believe, was really against it." Considering Stagg's frequent references to Dunstable's (and especially Petterssen's) excessive "gloom," the word "perhaps" does not seem to contribute much to clarification.

As midnight approached, with June 5 only forty-eight hours away, I felt we stood face to face with a crisis. Surely, I thought, if D-Day had to be postponed on account of bad weather, sea, and surf, the supreme commander ought to be warned now. Since at least May 31 we had been at a fork in the road. What we had done since was push the fork with us rather than choose one of the two possible paths. The next conference was scheduled for 3:00 A.M. I considered this to be our last opportunity; thereafter, my continued "gloom" for June 5 would serve no useful purpose. Fortunately this was not so. Only the ships in the northern ports were set in motion and they could be halted at any time within twenty-four hours. In other words, there were a few more hours available.

June 3 happened to be my day to speak for Dunstable. At the first conference I would have the opportunity to hear the views of Widewing and the Admiralty before my turn came. As I began to prepare, I recalled a piece of wisdom that a learned friend had given me many years earlier: The Chinese character for *crisis* consists of two elements—*danger* plus *opportunity*. Storm C (Fig. 19.1) was moving eastward, heading for Shetland; Storm D was following rapidly in its wake; and a new storm (E) was brewing somewhere to the east of the Great Lakes. Surely, I thought, a surge of rising pressure from the Azores was now out of the question. It seemed probable to me that opinions would now begin to converge, not toward the middle but toward the Dunstable "gloom." Perhaps opportunity was knocking on our door.

Widewing opened and repeated their belief in a protecting "bubble" of high pressure, an offshoot from the Azoric anticyclone. Though they conceded that low clouds might be bothersome in the western approaches to the Channel, the weather was considered "operational" for a few more days. With pressure rises over the Bay of Biscay and falls beginning to show up in Iceland, and with very few observations from the Atlantic, there was, obviously, much leeway for interpretations in favor of the views held by Widewing. Moreover, the High Command had planned for June 5 as the target date; enormous forces stood poised for the assault; naturally, one does not wish to oppose such plans unless one can see compelling reasons. The Admiralty, as they had done so often in the past, tried to steer a middle course with a leaning toward Dunstable, but an agreed forecast did not seem to be within sight.

While listening to those presentations, I felt we had reached a now-or-never situation. Not knowing that the forces in the north could be halted after being set in motion, I felt that we had no time to lose. So, I decided that it was right for Dunstable to take a firm stand. To make my point I referred briefly to the history of the large-scale situation since about May 30 and summarized the salient features as shown in the lowest section of Fig. 19.2. As usual, my emphasis was on the upper-air structures, expecting Douglas to supplement.

As I saw it, the strong and almost straight current aloft was unstable, in the sense that it would continue to produce rapidly growing disturbances. Cold air, coming straight out of the eastern part of the Canadian Arctic, was streaming southward and curving westward toward the Great Lakes, building up even stronger temperature contrasts to the west of the Grand Banks. Atlantic storms would continue to be steered east-northeastward. In the eastern part of the Atlantic details were less certain and it was difficult to say precisely where each storm center and weather front would be on June 5. My presentation continued with a general summary of the forecasts of visibility, clouds, weather, and winds as prepared by Douglas and his group: Strong westerly winds and low overcast would dominate the area of Overlord interest during the night of June 4–5. To emphasize the seriousness of the situation and to remove any impression of speculativeness, I ended by mentioning that the beginning of the change was imminent: Blacksod Point (on the west coast of Ireland) was already reporting strong winds and rapidly falling pressure.

The reaction to my presentation appears to have been varied. After the conference I was saddened to hear Douglas, this pillar of wisdom, if not of eloquence, say, "I am glad you were here to speak, for no one would have listened to me." Stagg, however, felt differently. Not even at this most critical time did he seem able to submerge his antipathies in the meteorology of the case. In his book (page 90), he writes, "As if he were embarking on an hour's lecture in a school of meteorology, Petterssen, representing Dunstable, treated the conference to a survey of the entire synoptic weather chart from the Great Plains of North America eastward across the Atlantic to Scandinavia and the Baltic Sea." Stagg continued, "As if to ram home his lesson, Petterssen told us that Blacksod Point . . . had just reported a force 6 wind [23–27 mph] and a rapidly falling barometer." As it happened, the "ramming home" of my lesson may have served a useful purpose for, according to Stagg, the Blacksod report had not yet found its way to the charts of other participants in the conference. Stagg, referring again to my presentation, continues: "On hearing it, in spite of his exasperating manner of presenting his thoughts, I think I would have reached over and shaken his hand if Petterssen had been on the other side of a conference table instead of about a hundred miles away at the end of a telephone." Though the mannerisms to which Stagg refers were mine, the substance of my presentation was intended to represent the Dunstable views.

At the 5:00 P.M. conference the Admiralty opened with a revealing analysis of the large-scale situation and drew attention to the new storm (E) that had formed to the east of the Great Lakes (Fig. 19.2) and was moving east-northeastward with great speed; it would soon dominate the Atlantic situation. In retrospect this analysis must be ranked among the most significant contributions to the conference that preceded the choice of D-Day. Nevertheless, on pages 55–56 of his book Stagg writes:

> . . . the naval forecasters seldom seemed to have a logically developed or distinctive opinion of their own. . . their participation was seldom decisive, often unhelpful and sometimes a little vexatious. Having no long-range prediction techniques of their own—if they had their nature was never divulged—and lacking the practical experience and theoretical background of the Dunstable participants, their support meant little if given to Dunstable and tended to make Col. Krick even less open to persuasion when it leaned toward Widewing.

Fig. 19.2: Surface weather maps for June 4, 1300 GMT and June 6, 0600 GMT.

During the morning hours of June 4 there was an enormous reorganization of the weather systems in the north Atlantic. A new strom F had absorbed storms C and D (see Fig. 19.1). At the same time storm E became dominant in the western Atlantic.

The upper map (June 4, at 1300 hrs) is of particular interest because it was mainly due to this map that the meteorologists forecast barely usable weather for the night of June 5–6. This map is also interesting because it shows the strong winds the forces would have encountered if D-Day had not been postponed.

The lower map refers to the situation on June 6 at 0600 GMT, which is shortly after the Allied landing on the beaches of Normandy. A rare "almost / just barely" weather situation.

Speaking for Dunstable, I could only refer to my early-morning discussion and add details revealed by later charts. In their broad aspects, things were going as expected. I supported the Admiralty views and added emphasis by referring to the enormous amount of energy input to Storm E, brought about by the deep mass of arctic air that streamed southward from the Canadian archipelago. Unbelievable as it may seem, Stagg (p. 94) states that I had the arctic air coming out of Russia and flowing into Canada. He also states that I predicted that the dew-point temperature in the Channel area would be the same as that at Bermuda, and that Storm E would be near Scotland on June 7. With reference to the weather situation at the time of the conference, Stagg quotes me as having said, "The situation has verified itself," a type of statement that I most certainly would not use. It should be noted once more that the quotation marks used by Stagg represent extracts from his own notebooks and do not necessarily reflect the thoughts expressed by the person mentioned. My presentation ended with a general summary of visibility, clouds, weather, and winds prepared by Douglas and his group: mainly low overcast and strong winds. Though the winds indicated were somewhat lower than those predicted earlier, they, as well as the clouds and visibility, placed the situation in the "nonoperational" category. Widewing, however, found new evidence of a pressure surge toward the Channel area and foresaw light winds and favorable weather for June 5.

In the late afternoon of June 3 the conference situation may be summarized as follows: The Admiralty and Dunstable were in substantial agreement, and some further discussion between them would, almost certainly, have eliminated residual differences. Furthermore, the chairman, though he did not choose to take the forecasters into his confidence, maintains that he, independently, had come to the conclusion that the weather conditions would be unacceptable on June 5. Yet no attempt was made to explore in depth the difference between Widewing on the one side and the chairman, the Admiralty, and Dunstable on the other.

Even at this late hour, could not the six forecasters have met, say, at Widewing, to try to reach an agreement, first on their analyses and then on their prognoses? Would it not have been possible for the chairman, or his deputy, or both, to attend? All concerned were within about fifty miles of Widewing. One may also ask, Was Sir Nelson or his deputy, or any other competent authority, ever informed that the conference mechanism had proved incapable of producing agreed forecasts? Was there really nothing that could have been done on June 3, or earlier, except "keep the ideas fluid" and "maintain a fine balance until the last moment?"

The prerequisites for a rapprochement were sadly lacking. Stagg, under the heading of "June 3," writes that he "was sometimes all but physically nauseated at the thought of a weather chart. . . ." And after the afternoon conference on the same day: "But I had not recovered from the afternoon's spate of forecasting jargon before I had to start preparing for the evening discussion at 8:00 P.M." In reading these outbursts, which do not reflect favorably on the profession of forecasters, one must sympathize with a certain Frenchman who tried to escape from someone's supercilious attitude by praying, "Save us, dear Lord, from the *connoisseurs qui n'ont pas de connaissance*, and from the *amateurs qui n'ont pas d'amour!*"[81]

The evening conference proved to be as infertile as its predecessors. The Admiralty and Dunstable were in broad agreement; Widewing was optimistic, though perhaps less so than earlier; and the chairman kept his meteorology and tactics to himself.

After a short nap in my office bunk I returned to work; Douglas had gone home for a brief rest. The next conference was scheduled for 3:00 A.M. on June 4. About 11:00 P.M. the Admiralty phoned and suggested that we discuss the situation. We exchanged a good deal of useful information and neither pressed the other for any kind of agreement; clearly, we were thinking along similar lines. The call was indeed a pleasant surprise; it was a very long time since anything so nice had happened to me.

But a greater surprise was to follow: Irv (Krick) phoned to hear if I could see my way to a more optimistic forecast for June 5. He and Ben (Holzman) had examined all June analogs for the forty-year series of charts at their disposal and found nothing to indicate that their views were wrong. Though time was pressing we exchanged a few words about our friends in Pasadena and the pleasant experiences we had had during my visit there in 1935. As to the present weather situation, I explained that nothing had happened that could cause me to become more optimistic for the night of June 4–5. Moreover, Douglas was resting, and in no circumstances would I change the Dunstable forecast without discussing the amendment with him. It was clear that Irv was concerned about the responsibility that rested on us and I am sure we both wished it had been possible to "split the difference." Three hours later the final and fateful decision would have to be made.

The Admiralty opened the 3:00 A.M. discussion and stressed that the situation had deteriorated further; they foresaw unsuitable conditions and predicted winds of force 6 (23–27 mph) at the time of the planned crossing. Their forecast was very much to my liking. I was next in line. Having heard the Admiralty's forecast and having had the benefit of the private discussions with the Admiralty and Widewing, I felt it wise to say very little, except to express agreement with the Admiralty and to invite attention to the rapid eastward movement of the storms on the Atlantic. Referring to my contribution, Stagg writes: "Petterssen on the other hand was cheerful because 'the situation is developing exactly as expected . . . all according to plan. . .'". Stagg, who did not know of the personal exchanges that had preceded the conference, appears to have obtained the impression that I had come to this all-important conference in a cocky mood.

Widewing did not subscribe to all the pessimism expressed by the other centers. Their analogs favored a new surge of high pressure that would provide some protection to the Channel area. The Admiralty and Dunstable remained unyielding and, according to Stagg (p. 100), Widewing informed him that they would leave the majority views unopposed. Considering the surrounding circumstances, this was a generous and most helpful contribution.

At 4:14 that morning Stagg briefed the supreme commander's conference. The forces already set in motion were halted and operation Overlord was definitely postponed, provisionally for twenty-four hours.

In the early hours of June 4 a sudden and major reorganization of the atmosphere over the Atlantic sector threw the forecasters into confusion. Much effort was expended in trying to find out what had happened and what was going on. Nevertheless, toward the end of the day, the three teams—the Admiralty, Widewing, and Dunstable—reached a state of harmony that had hardly ever been attained since February when conference discussions began. A new and sudden development in the mid-Atlantic had resulted in a major storm F (Fig. 19.2) which had absorbed Storms C and D and, possibly, a minor but potentially powerful disturbance, the identity of which could not be fully documented with the sparse data at our disposal. Also, Storm E, which had continued to grow rapidly, had begun to slow down in its eastward movement. In the upper atmosphere, the strong and almost straight current that had dominated the scene the whole week had suddenly changed into a wave-shaped pattern, and the amplitudes were growing rapidly. Both storms E and F had now acquired generous inputs of fresh energy through deep intrusions of real arctic air. Clearly, we were now in for a new type of large-scale weather regime.

The evening conference at 7:30 P.M. was especially productive. Though the conditions were deemed to be marginal, all agreed that the crossing of the Channel, with landing at dawn on June 6, would be possible. Residual differences were now small and unimportant. The forecasters' telephone discussions began at 3:00 A.M., June 5. Most of the uncertainties that had plagued us during the day had now been removed. In line with my policy of firmness in the Dunstable presentations, I stated that "it is now clear that Storm E cannot continue to move eastward as fast as I have predicted earlier." The arguments that had applied to a straight current aloft did not apply to a wavy current of growing amplitudes. All seemed to agree that the winds of Storm E would not affect the Overlord area. Uncertainties as to the behavior of Storm F remained, but in any case there would be a break between the two storms.

The High Command met at 4:15 A.M. June 5, while a howling gale was raging over the British Isles, the Channel, and northern France. Montgomery, who twenty-four hours earlier had been in favor of launching the assault with landing at dawn on June 5, writes in his memoirs that the postponement had saved them from a disaster. With the storm reminding all of the gravity of the situation, the supreme commander, between 4:15 and 4:30 A.M., heard all evidence and gave orders to launch the assault, with landing at dawn on June 6.

The forecasts that served to launch the assault were unique in that the methods used by the different forecasters varied a great deal, with but little impact upon the end results. Since Douglas and our colleagues in the Admiralty seemed "to have a feel" for the situation, some extrapolation, based upon experience, led to convincing results. What they foresaw would happen at sea level was in harmony with what could be expected from the upper-air analyses. Widewing, on the other hand, using their analog techniques, found that a surge of high pressure from the Azores would protect the western approaches. This, too, agreed well with the evidence of stagnation of Storms F and E, and growing amplitudes of the upper current. All six forecasters had arrived at perfectly compatible results, using different methods and expressing themselves in different terms. My personal contribution to the forecast for June 6 was to support and agree rather than to propose and argue. I do not think that any of the forecast centers can claim to have played a dominant part in reaching "the blessed end."

The question of the useful range of forecasts outlived the Overlord operation and is still active in some quarters. Those who consider a long-range forecast to be a continuous chain of detailed statements—something like a series of short-range forecasts—may well maintain that such forecasts for periods in excess of thirty-six or forty-eight hours are not possible, except (as is often said) in exceptional cases. On the other hand, those of us who consider a long-range forecast as a statement of the *predominant* weather over a specified period of time, beginning at the end of the interval to which the short-range forecast refers, will agree that forecasts for periods longer than forty-eight hours are possible, though not necessarily on all occasions. In the case of Overlord, the occurrence of a stormy period affecting the days of interest was foreshadowed as early as May 28 and consistently forecast on and after May 31. What turned out to tax our skills to the utmost was to forecast much in advance the short-lived peaks and troughs within this stormy period.

We must not forget to mention that the German meteorologists shared in the Dunstable "gloom." After the war's end we heard that they had predicted a stormy spell centered on the favorable tidal period. What they and the German High Command failed to appreciate was how much could be accomplished by the Allied forces during a very short break between major storms.

CHAPTER 20

Overlord: A No-Name Policy

Take but degree away, untune that string
And, hark, what discord follows!
—Shakespeare, *Troilus & Creddisa* I, iii

Our work and worries did not end on June 6. The weather remained marginal and the conferences continued. Forces, supplies, artificial harbor components, and gear of various kinds had to be shipped across the Channel. The air forces had to keep pounding Nazi installations as circumstances dictated and opportunity served. The beachhead had to be expanded before the expeditionary force could operate in a self-contained manner. However, on June 6 the conference tensions subsided and the discussions became more factual—and relaxed. Since longer-range forecasts were now of less interest, the uncertainties in the analyses across the Atlantic became less bothersome.

On June 9 the assistant director in charge of the Dunstable complex received a handwritten note from the deputy director, Mr. E. Gold, expressing his gratitude for the Dunstable contribution to the Overlord operation. The note was handed to me just as I was trying to get ready for another conference. I scanned through it, saw that no action was required, and passed it on to Douglas. I intended to obtain a copy for myself, but in the turmoil of work the episode was soon forgotten.

Though I did not read the note properly, I remember one phrase clearly: "A major disaster has been avoided." I was very grateful for this heartwarming message from Mr. Gold, a scholarly person, a versatile scientist and administrator who ruled with firmness over all weather services and was known to be somewhat conservative in his expressions of praise. A few days later I received an equally warm message from the director.

Sir Nelson to Dr. Petterssen, 13 June 1944

I am writing to thank you for the contribution which you made towards the Allied cause in connection with the invasion of Northwest Europe. The Meteorological Office was exceptionally fortunate in having your great knowledge and wide experience at its disposal, and it was a great relief to me to know that your clear vision and sound judgement were available for the task of framing the advice to be given to the Supreme Commander. I enclose a copy of a Minute which Group Captain Stagg has sent me expressing his gratitude for the help which he received from you. To this I would add that, not only am I very grateful to you for the support you gave to Stagg, but I appreciate even more the disinterested loyalty which you have shown, not only in connection with this operation, but ever since you first came to help me.

Now that D-Day has gone, I trust you will find it possible to ease up and enjoy some respite from the long hours and almost continuous strain to which you have been subject for several months.

The Minute from Stagg to which Sir Nelson referred reads in full:

The Director,
Meteorological Office (DMO),
Air Ministry, Kingsway, W.C.2.

Now that the first and most important decision in the assault on Europe from the West has been made, I would like to express through you my gratitude for the help given me by the Air Ministry Meteorological Office, and most particularly for the contributions from Mr. Douglas and Dr. Petterssen to the final meteorological advice given to the Supreme Commander and his Commanders-in-Chief.

As you know, the synoptic situation about the time when the decision for the scheduled H-hour had to be taken was unusually complex. Low pressure systems of an intensity, rate of development, and speed of movement appropriate to midwinter dominated the situation and uncertainties in the positions of these systems and their associated fronts contributed substantially to the difficulties.

It was therefore not surprising that over the frequent meteorological conferences by telephone which preceded all of the meetings with the Supreme Commander and his Commanders-in-Chief, there was but little or no agreement in the forecast advice offered separately by the three Centres.

It was then that the fine combination of true insight and long experience behind both Mr. Douglas and Dr. Petterssen showed up against all the text book analyses and arguments.

Insofar as the best advice was actually given to the Supreme Commander at the most critical times, that advice was based on Dunstable's views; it would not be exaggeration to say that those views were either actively opposed or only very half-heartedly accepted by the other participants in the meteorological conference even at the most crucial times.

I would be very glad if you could convey to Mr. Douglas and Dr. Petterssen my sincere thanks for their help in a most difficult situation. It must have been a very considerable strain for them to take part in so many conferences at such awkward times of day; a strain not lessened by their having to listen to, and take part in, discussions with which they could have very little sympathy.

Whatever other result their participation in the conference may have had on the success of the assault in Normandy in June, 1944, they have clearly vindicated their early contentions that except in unusual situations forecasting in this country for even 12 hours ahead is still a job requiring the utmost skill, caution and experience.

(Sgd.) J. M. Stagg
Chief Meteorological Officer.
Meteorological Section,
SHAEF
12 June 1944.

I was immensely pleased, and lost no time in sharing the good news with my staff. I was glad Stagg had been helped in a task that, to me, had seemed hopeless from the moment it was planned. Time and again during the weather conferences I had had to take a firm stand against what I considered to be undue optimism and indecision. However, somehow, the "blessed end" had been reached—and what else could matter? Who could now entertain grudges or nourish swollen pride?

Above all, I was glad that Sir Nelson was relieved. He had joined the Meteorological Office, as its director, in 1938, when the war was on the doorstep and the state of preparedness in forecasting was at a low ebb. Many of the things he desired could not be done overnight; he had to expand and disperse his services without having the opportunity first to put his house in order. Normally, he visited his home only every second weekend; otherwise, he lived and worked in his office. Overburdened with work and care, he had already begun to show signs of declining health. I felt that he, far more than I, needed "to ease up and enjoy some respite." Sir Nelson was one of the most conscientious and thoughtful men I have ever known. I was glad to have been of some help. I remembered our early discussion of the Overlord arrangements, and I was stirred by his mention of my "disinterested loyalty."

Late in the evening, when I had time to read the letters again, I was saddened to note the harsh tone of Stagg's minute. Surely, Douglas and I would not feel better thanked by reading derogatory remarks about our colleagues in the Admiralty and at Widewing. Not being a forecaster, Stagg was not in a good position to judge. As chairman of the conferences, he had done but little to provide the cement that was needed to keep the three teams together. Also, there was much more work to be done. If there were any hatchets about, they ought now to be buried.

Next morning, with very little time to spare, I wrote Sir Nelson a hasty note, thanking him for his warm message and unvarying kindness. I went on to suggest that it might not be wise to mention contributions by individuals, as Stagg had done, in connection with the forecasting for Overlord. It ought to be satisfactory for all concerned just to have it said that the meteorological profession had stood up to the test and rendered a signal service in the liberation of Europe. The crux of the matter had been the postponement of D-Day from June 5 to June 6. Historians might well agree that this was one of the several hinges on which the gate to freedom swung open.

As things calmed down, the supreme commander found time to sign letters expressing his gratitude to each of those who had contributed to the meteorological conferences and the happy outcome—a truly remarkable manifestation of the memory of his heart (Fig. 20.1). Since I had been invited to assist the British Service, the British Air Council took note of the supreme commander's letter to me and added their own congratulatory message.

Almost immediately after D-Day, extravagant rumors reached Washington and soon spread from coast to coast, mainly through newspapers and magazines. Though the frills varied within generous limits, the hard core of the rumors was the same: The British teams had failed in their predictions for D-Day, and Overlord had been saved by the U.S. Army Air Corps team, using Krick's analog techniques. Unlike old soldiers, these rumors never faded away. They are still active, and it stands to reason that, in the aggregate, their advertising value must have been

Supreme Headquarters
ALLIED EXPEDITIONARY FORCE
Office of the Supreme Commander

19 September 1944

Dear Dr. Pettersen:

I desire to commend you for your part in the coordination of the operation of the Meteorological Service in support of the 'OVERLORD' assault of the Continent of Europe. Considerable research and long hours of work by you and your associates resulted in the reconciliation of differences in forecasting methods and the development of a procedure which enabled me to receive the advice necessary for the selection of D-day with confidence that the information received was the best obtainable.

This service on your part, and those associated with you is sincerely appreciated and merits very special commendation as an outstanding contribution to the success of the Allied invasion.

Sincerely
Dwight D. Eisenhower

Dr. Sverre Petterssen
Meteorological Office
Air Ministry
London, W.C.2.

Fig. 20.1: Letter of commendation from General Eisenhower.

considerable, especially for Dr. Krick, who soon became heavily engaged in a variety of commercial consultant services, including forecasting for different time spans.

Some American commercial meteorologists, being concerned about professional ethics and fair competition, have tried to trace these rumors and claims to a common source, but all such attempts have failed to identify persons other than reporters and publicity agents. It is generally understood that their legitimate business is to find "wonderfulness" even where none exists. Reasonably enough, Krick could not be held responsible for distortions that appeared in articles not written by himself.

Eleven years after Overlord, Krick wrote his own account in a book addressed to the general public.[82] Compared with the claims and distortions that had appeared in American newspapers and magazines, his account is remarkably modest:

> The final D-Day forecast team was made up of six persons—two each from the British Air Ministry and the British Admiralty, and two from U.S. Strategic Air Force headquarters. The British, using short-range methods, could see no weather coming up in the unstable atmospheric conditions of those touch-and-go days of the first week in June, 1944, that would justify the risk of committing the great expedition to the stormy Channel crossing. The Americans, using analog forecast methods, saw "possible" weather. They also foresaw continued worsening weather for some weeks, if this present opportunity were passed up.
>
> General Eisenhower probably has never known how bitter was the division and argument within the forecast team around June 1 to 4, as to whether the next few days would provide any weather opportunity at all for the invasion. But at last the decisive word was taken to him that the weather should be "possible," and he ordered the great movement to start. In *Crusade in Europe* he tells how a great wave of relief and optimism swept all commands, irked and chafing from the days of sitting in bitter uncertainty because of the weather. A forecast ruling the weather "impossible" might well have delayed for a year the ending of the war.
>
> A final bit of irony is that Major Lettau, the chief German meteorologist, and his staff had agreed that the weather following June 4 would be much too bad to permit an invasion attempt, because of new storms moving in from the North Atlantic. As a result, the German high command had relaxed and many officers were on leave, and many troops on maneuvers. Poor meteorology on the German side, therefore, contributed to the success of D-Day, as well as modern methods on the side of the Allies.

The book was not well received by British meteorologists, and a review written by Stagg for the Royal Meteorological Society expressed his reaction in rather harsh terms.

In 1970, the United States could look back on a century of weather services. The beginning was made by the Army Signal Corps, but most of the time the U.S. Weather Bureau had occupied a central position. In the meantime, the air force and the navy had established their own meteorological organizations; so had many of the private airlines. An upsurge of commercial interest in meteorology that commenced after the war supported about sixty private enterprises in the field. Furthermore, in 1970, the American Meteorological Society could look, with pride, on fifty years of activity in the conservation and dissemination of knowledge and the advancement of professional ideals. All these interests converged in arrangements for the celebration of "A Century of Weather Service," under the auspices of the Department of Commerce.

The celebration called for the publishing of an historical account of the achievements since 1870. Mr. Patrick Hughes was commissioned to write it.[83] The book, with official trimmings—including a preface by the secretary of commerce—contains a chapter on weather and war. Here we find, among other things, a remarkable account of how D-Day was selected. Thus, on page 89:

> Probably the most critical and complex weather problem of World War II was the Normandy invasion. There should be no prolonged period of high winds to produce swell heavy enough to hamper the landing craft. Pilots wanted clear weather, paratroopers cloudy skies to protect them from German planes. The army, fearing a gas attack, wanted onshore winds; the navy wanted small waves, thus offshore winds. In addition, the landings had to be made at low tide, and the Allies needed at least three good weather days to bring equipment and supplies ashore after D-Day. If the weather conditions were wrong, the invasion might have to be put off for more than a year.
>
> A joint meteorological staff had been established in April, consisting of weathermen from the British Air Ministry and Admiralty and from the U.S. Air Forces weather service. The three groups were quartered separately so a single German bomb could not wipe out the entire weather staff. After choosing conditions that seemed best for all involved, the high command asked the meteorologists to calculate the climatological chances of getting the desired weather. The odds *against* it were 24 to 1 for May; 13 to 1 for June; and 50 to 1 for July. June 5 was tentatively chosen.
>
> By June 3 it was obvious the weather would not be good enough, so Eisenhower postponed the invasion until the 6th. On the evening of the 4th, the weathermen predicted relatively good weather for the 6th, and the signal was given to go. The next day, the last chance for cancellation, the American team still said go, one British team said no. The other British group was undecided; when it joined the Americans, Eisenhower was briefed and the invasion was on.

It is interesting to note that Krick, in his book, makes no reference to the forecasts that led to the postponement of D-Day from June 5 to June 6. Hughes, on the other hand, lets the American team predict not only the need for postponement but also the possible conditions on June 6. He also gives credit to one of the two British teams (the Admiralty) for having made a last-minute volte-face to side with the Americans.

In reading Hughes's story one is led to wonder whether there exists in the worthy science of topology a theorem that defines an upper limit to the amount of distortion at which identity is completely lost. The moral of Hughes's story might well be this: If you wish a rumor or a fantasy to be shrunk, don't take it anywhere near Washington, D.C.

A remarkable account of meteorological successes in World War II has been given by Theodore von Kármán, who repeats, endorses, and adds drama to the American claims.[84] We are told that it was von Kármán who discovered Krick, brought him to the notice of Millikan, and helped to get him established at Caltech to teach "reasonable meteorology." We are told, also, that during the war Krick not only picked the time for Eisenhower's launching of Operation Overlord but also chose the time for the Battle of the Bulge, the crossing of the Rhine, and other major operations. We learn, without identification of authority, that by these meteorological predictions, the advances into Germany of the forces under Eisenhower's command was accelerated by two months.

In reviewing these and many other accounts of the meteorological aspects of the Overlord operation, I have come to the conclusion that much of what appears to be outright distortion is really innocent repetition of rumors that have circulated for so long that they have acquired a patina of veracity. In John Knox's summary of Arth's *Discourse on Cursing* we are told that "Veritie is the strongest of all things."[85] Although, undoubtedly, this is so, one must realize that other things may be more newsworthy.

Though many of the claims, distortions, and suppressions that have appeared in newspapers, magazines, and books have been annoying, I have felt that my no-name policy was sound as long as no one descended to the level of personal derogations and unfavorable reflections on the profession of forecasters. What caused me—thirty years after Overlord—to discard my policy was not the proverbial last straw: It was Dr. Stagg's book.

We know very well that, in dealing with a complex system like our atmosphere, when one has to reduce theory to practical procedures and, from there, go on to predict future events, great latitude is required. The adjustments, to be useful, must be based on experience gained from one's handling of past cases of similar structures and histories. What other source of guidance can one think of, except, perhaps, divine inspiration? But when someone, however skilled in the arts and maneuvers of management, does not know the theories to be adapted, and has no experience that could be called upon for this purpose, had he possessed knowledge of the theories that might be applied, and when he, among other things, uses a popular book to question the scientific competence and professional skills of his helpers, then one arrives at the conclusion that the no-name policy has outlived its usefulness.

CHAPTER 21

Washington and Honolulu

Si fueris alibi, vivito sicut ibi.
[If you are elsewhere, live as they live elsewhere.]
—St. Ambrose

As the Normandy beachhead expanded and facilities and services were established behind the lines, the pressures on Douglas and myself decreased greatly. The meteorological part of Overlord had come to an end. Although our advice was sought on occasion, we had no direct responsibility for specific operations. At the same time, work in the Upper-Air Branch changed noticeably. Since long-range night bombing had been largely replaced by short-range day bombing with fighter escort, our involvement in bombing operations became far less demanding. For me, the term "spare time" had again acquired meaning.

Meanwhile, as the Americans were preparing for their assaults on Japan, it was clear that a great deal of bombing had to be done before landings could be attempted. The always alert U.S. Navy felt that not enough had been done to gain wisdom from the meteorological lessons learned from the bombing of Germany. Though Washington had heard a great deal about the new analog techniques used by the army air corps, the navy reserved their judgment. Their job was to please Admiral King, and the army air corps had to satisfy General Arnold. The judgments at the summit differed on occasion.

In late August 1944 I received an informal invitation to come to Washington to conduct a series of seminars on the work in my Upper-Air Branch and its use in the bombing of Germany. To satisfy protocol, it was arranged that my presentations should be held in the U.S. Weather Bureau, while officers in the army air corps and the navy would be invited to attend. Other formalities complicated the arrangements. As a Norwegian officer I could receive no kind of compensation from a "foreign power." The easy way out was for the U.S. Navy to consider me as a guest of Admiral King and to pay my expenses out of his hospitality funds. Commander Stoars, one of my early MIT students, was assigned to attend to my needs, to take me wherever I wanted to go, and to pay my bills—indeed, a most comfortable arrangement.

It was understood that at least one of my lectures would discuss the Overlord business, and I had prepared well so as not to uncover controversies. However, when I called on high Weather Bureau officials and happened to mention Overlord, the conversation was soon diverted to some other subject. I was rather surprised that the meteorological community in Washington showed so little interest in a matter that, to me, ranked first and foremost in the history of weather forecasting. After the visit had come to an end, I remarked on the lack of interest to my well-informed naval friend. He explained that the Weather Bureau officials had only tried to be polite. Everyone knew, and many believed, the rumors in circulation: The U.S. Army Air Corps team, using analog techniques, had saved the Overlord operation from disaster, while the British team—Douglas and myself—had excelled in bungling the case. Although I decided to eliminate Overlord from my seminar program, I was very pleased to accept an invitation to a small evening gathering of naval friends to review certain aspects of the cross-Channel operation.

Though my Washington seminars were well attended, I felt that I was not talking to the right audience. My naval friend, who had initiated my invitation, then made arrangements for me to go to Hawaii to talk to a group of air force and naval meteorological officers, some of whom would be directly involved in the assaults on Japan. In Hawaii I met a keen and sincere group. Some of the senior naval officers had served on aircraft carriers under Admiral Halsey's command and had had to satisfy him without having much data on which to base their forecasts. Halsey, I understood, was greatly admired by those who had served under him. Able men seem to prefer a demanding and just commander.

The group of air force meteorological officers was led by Colonel (later General) Bill Strong, a leader of exceptional skill and firmness. He saw to it that the discussions remained centered on the primary problem: how to forecast for the bombing of the large cities in southern Japan. In one of our sessions I pointed out that, in the winter, they would have to be prepared for westerly winds of about 250 mph at bombing levels. In such cases, the heavily loaded planes would have to fly at low levels west of Japan, then rise and ride eastward on a strong tailwind.

Since my remark did not provoke any comment or discussion I feared that I, naïvely, had offered them "another glance at the obvious." Far from it. As I was told much later, no one in the room had ever heard of such winds. They had felt that they were listening to a "college professor" who had come with some fancy theory but had no practical experience. The silence that followed my remark had been nothing but a reflection of their politeness. They were all very kind and hospitable men.

In fact, hospitality became one of my major problems. Normally, I would work from 9:00 to 12:00, starting with a lecture and continuing with a discussion or question-and-answer session. At about 12:30 I would be taken to lunch, often in the Outrigger Club overlooking Waikiki Beach. Thereafter, boating, riding, sightseeing, cocktail parties, and dinner, followed by after-dinner gatherings in different places. For reasons I did not understand, a curfew started every evening at 10:00, but since I was taken about by high-ranking officers, the gate was always open to us.

One day I let myself be taken out on the firm understanding that I would be returned to my quarters before curfew time, the reason being that I was determined to have a full night's sleep. My companions did their best, but traffic was heavy and we arrived at the gate a few minutes late, together with many stragglers. The poor guard could make no exception, and when my senior companion referred to his rank and started an argument, the guard fined each of us ten dollars (as he had a right to do) instead of ten dollars for the whole car (as he might have done). The situation became overheated, and no one would listen to my offer of paying forty dollars, ten for each of the four of us.

Next morning my senior companion took the case to the provost marshal. His ruling impressed me immensely. One, since I was a foreign officer and a guest, the U.S. government had no authority to impose any fine on me; two, since X and Y were officers "in line of command," it was their duty to entertain me, so no fine was in order; three, since Z was a medical officer in charge of a hospital section and had no business to be in my company, he had to pay ten dollars.

I left Hawaii in late October, stopping off in Los Angeles to see my friends Jack Bjerknes and Harald Sverdrup. All three of us wanted to return to Norway after the war, provided that we could do useful work without blocking the promotion of deserving men who had remained in Norway during the occupation. The sense of nostalgia that came from our conversation made a very deep impression on me.

I returned to Dunstable in late October. In mid-January I received a radio message via the American Embassy and the British Air Ministry. The message, which came from some commanding general in the Far East, had obviously been initiated by General Strong. It just said that on a recent bombing mission to Japan the force had encountered 247 mph westerly winds at bombing level. It was only long after the receipt of this message that I was told of the response to my mention of 250 mph. Actually there was no great skill involved on my part. We had observed winds of about 200 mph at high levels over England; all I did was to add 50 mph to allow for the stronger temperature contrasts along the east coast of Asia.

CHAPTER 22

Tea in the House of Lords

These little things are great to little man.
—Oliver Goldsmith, *The Traveller* (1764)

Toward the end of the war, as the Russians began their advance into Norwegian Lapland, the Germans withdrew in good order, carrying out scorched-earth operations of extreme severity: No house, shack, or shed was left standing. In this area of inhospitable climate, where the population density was less than three persons per square mile, reconstruction after the war would be exceedingly difficult. The continued decimation of the reindeer herds might have serious consequences for the nomadic part of the population.

The state of affairs in Norwegian Lapland became a matter of great political concern. A few went so far as to express doubt that Norway would ever regain control of the province. Others, holding more moderate views, feared that the Norwegian Underground might find it difficult to understand why nothing could be done to hinder the German retreat and destruction, and limit the Russian advance. A few thought that Norway could "go it alone," withdraw her units from the British forces and attack the Germans in their rear. Effectively, this would mean reoccupying the strip from the mouth of the Lyngen Fjord to the Swedish border. The advocates of this scheme had developed an escape clause: Even if the military operation should turn out to be a failure, the political gains would justify the enterprise. Only inaction was considered unacceptable.

Though the political aspects were far from clear and the forces behind the scheme were hard to identify, it appeared that the general idea—that something ought to be done—had considerable support. However, the withdrawing of Norwegian units from British overlordship was no simple matter and could hardly be handled through customary channels. Nevertheless, one or two men of consequence thought it might be useful to find out informally what the British reaction might be should things go from bad to worse. When my opinion was asked, I suggested that the most useful contact might be Lord Cherwell, earlier known as Frederick Lindemann, professor of physics at Oxford. Cherwell held the sinecure position of paymaster general but his real power derived from his friendship of long standing with Churchill and his position as science adviser to the prime minister. I did not know Cherwell personally but I thought he could be reached through Sir Edward Appleton, who headed the Department of Scientific and Industrial Research.[86] I also suggested that a Norwegian physicist should be brought into the picture, since this would make informal conversation come more naturally. The outcome of these maneuverings was that Lord Cherwell invited Appleton, Dr. Olav Devik,[87] and myself to tea in the House of Lords.

The early conversation was centered on the German V-2 rockets, their range and potentialities. Then, at a suitable moment, Appleton brought up the Lapland problem, and Devik and I filled in with information and details. In the early twenties Devik had served as a meteorologist in the Tromsø Institute and both of us were intimately familiar with the Lapland region. I had heard rumors that Cherwell was an impatient man and did not gladly suffer his time being wasted. Nevertheless, he seemed quite interested in the Lapland problem. He asked a few questions and jotted down a few notes. Appleton and Cherwell were also interested in problems unrelated to Lapland. I gained the impression that Appleton had fairly close connections with Clement Attlee, the deputy prime minister. Cherwell had several suggestions for items that Attlee might wish to include in a speech that was about two weeks off. Likewise, Appleton had several wishes concerning a speech that Churchill was preparing. No wonder that top politicians often seem to be well informed.

I found this horse trading very interesting, and I began to wonder whether our Lapland problem had provided a welcome opportunity for Appleton and Cherwell to transact business without formalities. As we broke up, Lord Cherwell said that he might get in touch with us soon. A few days later we were told, in rather guarded terms, that "it was thought" that the German retreat in Lapland was about to end. It might even have stopped already, and no important developments were likely within the foreseeable future. This soothing message turned out to be completely accurate.

CHAPTER 23

Reconstruction and Travel

As I brew I must drink.
—Proverb

The compactness of the Norwegian headquarters in London ensured efficient communication and circulation of official information as well as of rumors and gossip. Though there was nothing secret about my visit to Washington and Honolulu, it was thought unwise to have anyone believe that bombing of Japan was on the agenda. On a few occasions when an explanation seemed necessary, I let it be known that I was going back to the United States for a week or two to renew contacts. This seemingly innocent stratagem, however, started new rumors and soon led to unexpected results. Riiser-Larsen, who considered himself responsible for having dislodged me from MIT, felt inspired to act once more. As a result of his good intentions, I received a letter from the secretary of education informing me that he had been authorized by the government to request that I return with them at the end of the war to assist in the reconstruction work, particularly in meteorology, aviation, and related fields. Though the government in exile could not make commitments on behalf of the Norwegian Parliament, it was their intention to propose the establishment of a suitable position for me if I would agree to return.

Since I received this rather extraordinary letter as I was getting ready to leave Washington, there was but little time for me to consider consequences and to compare it with certain American opportunities that had been under consideration for some time. Being guided more by sentiment and patriotism than by advice and shrewd calculations, I accepted the invitation on the understanding that I could do useful work without standing in the way of deserving men who had suffered the Nazi occupation of Norway. Though these understandings were rather loose, I used suitable opportunities during my visit to Washington to promote disengagement rather than engagement.

On May 8, 1945, the Nazi regime in Norway collapsed. The next day Riiser-Larsen was on the telephone, talking like an inspired poet—Riiser at his best. Eleven months earlier, in fact a few days after the postponement of D-Day, Riiser had told me that he would see to it that I got a seat on one of the first planes to return to Norway when the war ended; he had not forgotten his offer. I explained that it would take me at least two days to pack, make arrangements for my staff, and say goodbye to Sir Nelson and others; Riiser arranged to have a seat reserved for me on May 13.

On this spring Sunday, and in glorious weather, we flew in low over the Skaggerak coast, circled over small towns, saw flags flying from every pole, men waving, and women tearing linen off their clotheslines to signal their welcome—a nation gone hilariously mad with joy after 1,856 days of Nazi tyranny. The only thing our pilot could do was to rock the plane in mute response. We landed in the dismal atmosphere of Oslo airport, still serviced by *Luftwaffe* personnel clad in uniforms not very different from ours.

In Article 8 of his ultimatum to Norway, presented on April 9, 1940, Hitler had demanded that the Norwegian Weather Service be placed under the command of the *Luftwaffe*. It soon transpired that the Norwegians could not be relied upon to produce forecasts of such accuracy as the Nazis had expected. Our meteorologists in Norway had to spend five years in inactivity while forecasting techniques progressed in the occupied countries. The Germans had established an efficient telecommunication system for meteorological purposes, a fine net of upper-air sounding stations and forecasting units at all Norwegian airports. Sir Nelson had been quite right in saying, as we parted, that though Norway might be leading the world in meteorological theory, there was much practical rehabilitation work to be done to meet postwar demands; his representative in Norway had instructions to work very closely with me.

In Norway, science and scientific services, including weather forecasting, are controlled by the Ministry of Church and Education, which (its benevolent attitudes notwithstanding) proved sadly lacking in speed and funds. However, working out of the Royal Norwegian Air Force, and enjoying the wholehearted support of Riiser-Larsen, I managed to replace the German meteorological personnel with (sometimes hastily trained) Norwegians, rearrange the meteorological communication system, and maintain the upper-air sounding stations. As the leftover German supplies dwindled, friends in the American air forces in Europe helped out with balloons and other sounding gear. Eventually, it became illegal for the U.S. Army Air Corps to dispose of equipment in this manner. Although the Ministry of Church and Education recognized my needs, they were at the end of the queue waiting for funds. At this time, General Bassett, one of my 1935 students at Caltech, stepped in, loaded his personal plane with equipment, and flew repeatedly from Frankfurt to an air base near Oslo. All I had to do was arrange for the pickup. Eventually, regular funds became available for a somewhat reduced program—a program that some of the interested parties looked upon as having resulted from some kind of meteorological megalomania.

While I was struggling to replace *Luftwaffe* personnel and reorganize the weather service in Norway, international problems came to the fore and claimed a sizable part of my time. At its very inception (about 1855), weather forecasting became dependent on exchange of reports and other collaboration across national boundaries. In 1873 the International Meteorological Organization (IMO) was constituted as an informal body, more or less a club of directors of national weather services. With the advent of aviation came new demands. The needs for international collaboration increased further and reached hemispheric proportions in the late thirties. Then, as Hitler unleashed the horrors of war, this delicate structure of meteorological coordination crumbled; the collapse was unnecessarily severe.

Unfortunately, the allied countries had no meteorological Eisenhower to weld together diverse ambitions and forces into a productive whole. Traditionally, the United States had shown but little interest in international arrangements (except with Canada). As the war approached, the U.S. Weather Bureau had yielded to the military services, hoping that things would return to normal when the war should end. However, as is often the case, the point of no return proved to be at the very beginning. The term *normalcy* is a matter of definition.

During the occupation of France, anti-Pétain[88] meteorologists had gathered momentum, and emerged victorious when the war ended. France, with her long tradition in international collaboration, had to pause and seek a way through a labyrinth of new problems. The usual concern for prestige had become aggravated by the circumstance that the English language had become dominant at the postwar conference tables.

In the Soviet Union, able and forceful men had advanced to the helm, but their country, having been ravished by the Nazi hordes, lagged behind their allies in the development of new equipment and techniques. Mutual distrust and an *e-superiore-loco* attitude in the West tended to create a sort of "chicken-wire curtain" through which only small things were allowed to be exchanged with the Soviet Union.

All the real and imagined difficulties that stood in the way of resumption of international collaboration congealed into a formidable mass when, shortly after the end of hostilities, the president of the IMO became incapacitated. Thereafter, the problems were allowed to drift. The "paulo-post-future"[89] of international meteorology looked rather bleak. Being an unofficial organization, IMO was not well equipped to cope with the many problems that war had created. It was not until 1951 that it was revived, under the name of the World Meteorological Organization (WMO), and was recognized as a specialized agency of the United Nations.

In the meantime our friends in civil aviation had been very active. While the war was still in progress, the U.S. government, with the well-advertised blessing of Roosevelt, had taken the initiative to create PICAO—the Provisional International Civil Aviation Organization. It was understood that the *P* and *Provisional* would be dropped as soon as a peacetime convention could be ratified by all interested countries. When the war ended PICAO lost but little time in calling a conference at their headquarters in Montreal to set in motion work on meteorological arrangements for civil aviation.

Being overcommitted in other fields, I had not planned to attend the conference. However, events took a different course. An eager member of the hastily assembled PICAO staff had slipped in a conference document that opened thus: "In view of the inactivity of IMO, PICAO must. . . ." When Sir Nelson saw this rather extraordinary reference by one international body to a sister organization, he phoned me from London, asking if I could go to the Montreal conference as a Norwegian delegate and have the document withdrawn. When I suggested that I might not succeed, Sir Nelson continued: "Your influence is greater than you think; people consider you to be an impartial and unattached person, representing no one but yourself; you are thought of more or less as a stateless person living halfway between America and Europe. . . ."

So, with but little time for preparations, off I went, stopping in London to see Sir Nelson and continuing to Montreal, where I arrived during the opening address by a PICAO official whose identity was not clear to me. I was much impressed with his presentation and somewhat apprehensive of what was to follow. When the discussion of the agenda commenced I referred to a certain working document and proposed that the corresponding item be deleted from the agenda. The PICAO official in the chair seemed uneasy. However, after a brief discussion, a happy compromise was reached: The agenda item was reworded and the supporting paper withdrawn. I thought that I had gained my point, though perhaps at the cost of precious goodwill.

Early next morning I was awakened by the telephone: The president of PICAO invited me to breakfast at 8:00 A.M. At the appointed time the PICAO official who had presided over the opening session introduced himself: Edward Warner. I immediately started the mental process of preparing my defense. We sat down and chatted about things in general, the enormous potentialities of civil aviation that the war had created and the part that PICAO could play in furthering progress. Then Warner exploded his bomb: Would I join his staff to take charge of their meteorology program and guide it into a fruitful collaboration with IMO, without loss of ability to respond quickly and effectively to changing needs? I had two months to decide. Although I had to decline the offer, the breakfast meeting between Warner and myself led to a fruitful collaboration between us, much of which took place outside conference rooms.

Our first PICAO meeting had been concerned mainly (perhaps too much) with formal and constitutional problems. Soon, Warner called a second meeting to come to grips with basic aeronautical procedures. Being anxious not to become deeply involved in international aeronautics, I arranged to arrive twenty-four hours late, so as not to be present during the election session. This stratagem did not work. Since the election of a chairman had been postponed, I was left with no graceful means of escape; I was elected to serve as chairman of the meteorological division. In that capacity, I was a member of the steering committee of the main conference.

Though progress was satisfactory, it was rather slow. Traces of postwar suspicions were clearly in evidence. Precious conference time was often wasted on problems of translation. In an endeavor to create a more congenial atmosphere, I used a suitable opportunity to suggest that we take an evening off to dine together. This stratagem worked exceedingly well, mostly because I overlooked a translational finesse. To avoid any misunderstanding I had ended my suggestion by saying, "Of course, it will be a Dutch treat." Immediately, the head of the Netherlands delegation stood up, six foot five inches tall, and declared rather solemnly that he had no authority from his government to foot the bill. When my slangy expression was properly translated, loud and prolonged laughter followed. The dinner turned out to be a great success and the conference transactions accelerated noticeably. The conference achieved what it had set out to do: It developed a set of general procedures. Although many things have changed in later years, the basic framework still remains.

The most pressing problems in international aviation after the war were centered on the North Atlantic region. During the war our aviators had been limited only by that which was considered barely possible; risks, calculated or not, had to be faced. Now that the war had been won, the operations had to be scaled to that which was profitable and reasonably safe: *Payload* and *safety* had become matters of primary concern. With improvements in postwar aircraft, profitable traffic could be maintained between Gander, Newfoundland, and either Shannon, Ireland, or Prestwick, Scotland, but safety was marginal. Upper winds might exhaust the fuel supply and fogs might hinder landing. For each flight one had to determine a point of no return; before this point was reached one had to decide whether to continue or return. Should the weather turn bad on the takeoff field, the return might be a very risky affair.

To a large extent the economy and safety of transatlantic flights depended on efficient forecasting and reliable navigational aids and services. But since conditions were marginal, an ever-vigilant search-and-rescue service had to be provided. Strict and coordinated procedures for these related activities had to be provided under the auspices of PICAO and adhered to by all concerned. Under Warner's stimulating leadership, many of us worked hard to bring forth a new creation: transoceanic civil aviation on a commercial basis. With customary speed Warner called a regional conference to be held in Dublin in March 1946. Though its geographical domain was the North Atlantic, it was hoped that it would set a pattern for other regions to follow. I again served as chairman of the meteorological division and member of the conference steering committee.

With my attachment to upper-air techniques, for practical as well as for theoretical purposes, I found it difficult to visualize real progress in forecasting unless the oceans could be spanned by upper-air sounding stations. I saw the Dublin conference as a useful platform from which a "trial balloon" could be launched. My immediate concern was not to stroke "the international cat" the wrong way. Obviously, establishing an adequate network of sounding stations on the North Atlantic would require funds far beyond anything previously contemplated by the meteorological communities. Though international funds in the range of 50 to 100 million dollars a year might sound daunting, there were a few cornerstones on which I could build. 1) Patrick McTaggert-Cowan, the Canadian delegate, who had been in charge of the forecasting of Gander for the wartime Ferry Command [see chapter 7], could speak with greater authority than anyone else about the importance of upper-air winds for transatlantic flights; 2) since, as usual, the United States would have to absorb much of the cost, the support of their representative, my good friend Delbert Little, was essential; 3) since Warner would certainly lend support to any progressive scheme to further civil aviation, his influence with the telecommunication and search-and-rescue conferences would, perhaps, be decisive; and 4) Sir Nelson would surely be helpful if the Dublin conference could agree on a convincing recommendation.

I worked on this project largely outside the regular conference structure and checked now and then with Warner. Mac, Del, and I prepared a draft recommendation for the establishment of thirteen floating observation posts (Fig. 23.1) to be serviced by thirty ships especially equipped to provide regular soundings of the atmosphere up to great heights. These ships would also perform a number of other services for the benefit of meteorology and aviation.

The recommendation, which was passed unanimously by my division, received strong support from the PICAO Council. When the light, which had never been red, turned from amber to green, Warner, Del, Mac, and I purred like a hundred cats. However, to pilot a well-intentioned resolution through an international meeting is one thing, but to implement it—with hardware, cash, and services—is something else; no one could predict when the light might turn red.

As usual, Warner's drive proved irresistible. Six months after the Dublin meeting, delegates from the United States, Canada, and eight European nations with direct interests in transatlantic aviation gathered in London for the Conference of North Atlantic States on Ocean Weather Observation Stations on the North Atlantic. The conference, which was headed by Sir Nelson, divided itself into a financial and a technical commission, the latter being chaired by myself. A second wave of purring commenced when the plenary session agreed that thirteen

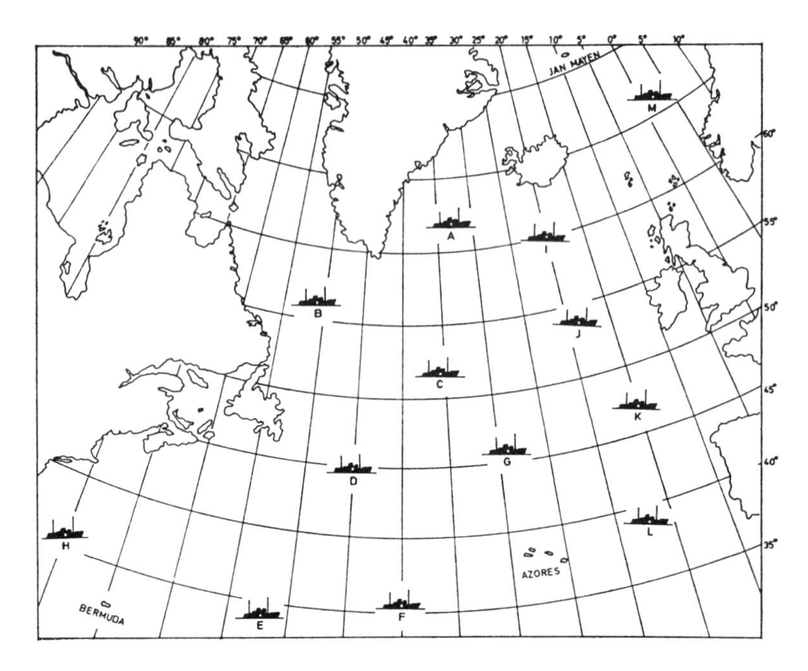

Fig. 23.1: PICAO array of thirteen floating observation posts established as a result of the conferences in Dublin (March) and London (September) of 1946.

stations were necessary in the interests of civil aviation. Before this stage was reached, the United Kingdom delegation had declared that in no circumstances would they consider anything less than twelve; two delegations had argued in favor of fourteen.

The financial problems were soon solved. Because of the many currency restrictions that followed in the wake of the war, all countries preferred to contribute in kind rather than to pay toward a central organization. The United States and Canada offered jointly to operate and pay for eight stations, France undertook to maintain one, Belgium and the Netherlands offered one, and the United Kingdom offered two. Station M, the thirteenth, became a problem child. Admittedly, it was of great interest to Norway, but since it was rather far away from the transatlantic air routes then contemplated, the costs would exceed what could reasonably be expected of Norway and Sweden. Very generously, Sir Nelson offered a United Kingdom contribution of thirty-five percent in addition to the two stations already mentioned. After some friendly bargaining, Sweden undertook to pay forty-three percent, and Norway was pleased to operate the station and pay twenty-two percent of the cost.

The work of the technical commission proved rather onerous. Not only had we to reach agreement on a variety of meteorological, navigational, and search-and-rescue procedures, we also had to keep the costs to a minimum. However, a general atmosphere of optimism and goodwill prevailed; the conference finished on time and the meteorologists present considered that major and lasting gains had been made. Dr. Warner, whose primary concern was civil aviation, was generous in his praise.

Dr. Warner to Dr. Petterssen 25 September 1946

I am writing to you, as head of the Norwegian delegation at the present conference, to make proper record of the immense gratification that everyone associated with PICAO will feel at the outcome of the meeting here; and to record the admiration that I have felt for the spirit in which the Conference has worked. There has been a harmony of purpose and eagerness to go to the very limit of possibility in cooperating to attain the desired result that should be recognized as noteworthy and as a bright augury for the future success of international arrangements for the support of air navigation services. It has been a memorable experience to have taken part in the discussions here, and I hope that you and your associates in the delegation and in your Government will share my satisfaction at the outcome.

Mr. Barringer, who had headed the large American delegation that had contributed so generously to our work, was equally pleased.

Mr. Barringer to Mr. Petterssen 27 September 1946

It is a pleasure to convey to you the gratitude of the entire membership of the Delegation of the United States of America to the PICAO Ocean Weather Stations Conference for the splendid manner in which you carried out your official duties as Chairman of the Technical Commission.

There is no doubt in the minds of any of the United States Delegation that the Conference was marked by a most sincere and enthusiastic desire to arrive at a workable agreement, and that your efforts contributed immeasurably to the success of the Conference.

My next problem was to find two suitable ships to serve Station M. Friends in the Royal Navy advised me that corvettes might become available, but the purchase cost was likely to be high, and alterations and refitting might be expensive. Where could a poor man find cheap ships capable of operating in the winter storms of the Norwegian Sea? Had I been too optimistic? I could see failure as a distinct possibility, and failure at this juncture might be hard to live with.

Never has a helping hand come so timely as on this occasion. Sir Nelson invited me to lunch with the undersecretary of the treasury. After a very pleasant meal, it was agreed that I could buy two corvettes at the nominal price of 10,000 pounds each. So off I went to inspect the ships, whereafter I signed the necessary papers—subject to approval in Oslo.

This additional manifestation of meteorological megalomania was fairly well received at home. And when the budget and the treaty concerning Station M came up for discussion in Parliament, the secretary of church and education, feeling unhappy at having to defend the purchase and operation of ships, invited me to sit by his side to whisper advice if the debate should take an awkward turn. The budget squeezed through, and the Ocean Weather Station program ceased to be a dream.

In later years there have been minor adjustments on the North Atlantic and extensions to the North Pacific. Without upper-air observations from these ocean areas, recent developments in forecasting with the aid of electronic computers would certainly not have been possible. It is in this general area that the future of weather forecasting lies.

Early in 1946 the president of IMO, who had suffered a severe accident soon after VE-Day, had recovered so much that the directors of the national meteorological services could meet to discuss postwar problems. The conference decided to relieve me of my duties as president of the maritime commission and, instead, to appoint me president of the aerological commission, the successor to the commission for the exploration of the upper atmosphere.

Traditionally, this was the most prestigious commission. My predecessors included such dignitaries as Vilhelm Bjerknes, Sir Napier Shaw, and similar men on the Continent. But now, after the war, practical problems overshadowed other considerations. It was not until August 1947 that my commission could meet. In the meantime Sheppard and I extended and formalized the work we had done in the Upper-Air Branch during the war [see chapter 10]. Through the approval of the commission and adoption by the conference of directors, our set of specifications, constants, functions, and various methods came into worldwide use.

Fig. 23.2: Petterssen and Eric Palmén discuss "upper-air" problems during IMO meetings in Toronto in 1947.

The Toronto meetings were truly remarkable in that all subordinate bodies of IMO met according to a staggered timetable. Never before had so many meteorologists, from every part of the world, met to exchange information and plan for the future. There I was reunited with my old friend, Eric Palmén (Fig. 23.2).

Eight years had passed since the last IMO meeting in Berlin in 1939. A number of new faces were seen and much new competence deserved recognition. Of the many newcomers, two men became of particular importance to me. One was General Donald N. Yates who, during the Overlord months, had served as deputy to Group Captain Stagg and now headed the USAF Weather Service. Until August 1947 we had not met in person. Soon, a long and lasting friendship followed. The other person of special interest was Dr. S. K. Banerji, director-general of observatories, India. Before 1947 drew to a close, I was on my way to a tour of India.

CHAPTER 24

At Crossroads in India

Jag ville jag vore ett drömlands son,
En infödd av Indialand.
[I wish I were a dreamland's son,
A native of Indialand.]
—Gustav Fröding, *Jag ville, jag vore*

Soon after the Toronto meetings Banerji inquired whether I would accept an invitation from the Indian Science Congress to attend their annual session, this time to be held in Patna, the capital of Bihar, in early January 1948. If I were able to accept this invitation I would also be invited to serve as adviser to the Indian Meteorological Service for a period of about ten weeks. I would be expected to visit a few of the major universities and meteorological centers and to give a series of lectures and laboratory instruction at the Central Observatory at Poona, where leading meteorologists would be present during my visit.

Before India had gained her independence, the Congress Party had stressed scientific advancement as one of their major political aims; Nehru himself had served as chairman of their political science committee. After the liberation, in August 1947, Nehru, now prime minister, had not lost interest. In an endeavor to stimulate science activities, funds had been made available to the science congress to invite a few foreign scientists to attend their annual session. Honorarium and expenses related to my work as adviser to the meteorological service would be paid by the Ministry of Transport. Needless to say I was very grateful for the opportunity to see India at this time of historic events.

On my way from Oslo to Calcutta I had occasion to observe the functioning of the aeronautical procedures that I had helped to develop under the auspices of PICAO. Foul weather kept us captive for twelve hours at Amsterdam, where an atmosphere of calm routine prevailed. Approaching Rome next day we ran into another storm, with much commotion at the airport. Our plane was allowed to land under two smaller aircraft that were circling in a deep cloud mass, hoping for a chance to land. Eventually, one of the planes found its way to a distant airport, while the other, with less fuel, continued circling. Next morning, when the weather had improved, we were further delayed until the wreckage had been examined and cleared from our runway.

We reached Tel Aviv on New Year's Eve, as the sun was setting. The war between Great Britain and Israel was on and distant shooting was heard, especially in the direction of Jerusalem. I tried to negotiate a tour of the Holy City in a British armored car, but after much bickering I was turned down. Security was not the only problem. Because of currency restrictions I had left Oslo three days before with only ten pounds in my pocket; after overnight stops in Amsterdam and Rome, there was not much left. Dead or alive, I might become a problem.

As 1948 dawned we took off, flew over Bethlehem, continued across the Dead Sea and the almost dead desert, refueled at Bahrain, and reached Karachi in the early afternoon. While the crew rested, the director of the Pakistan Meteorological Service entertained me generously and told me much of interest about the upheavals that had followed the partition of India into Hindustan and Pakistan. Millions of people from both states, dressed in rags, carrying little bundles and leading starved cattle, were on the move, mostly across the Gangetic Plains, to find new homes. The tolerance that had long existed between Muslims and Hindus had come to an end; Mahatma's preaching and his *satyagraha* (truth force) seemed to be of no avail. In the beginning Gandhi had taught that *God is truth*. More recently he had come to the conclusion that *truth is God*. In practice the difference seemed immaterial; the urge toward cruelty seemed to be shared equally by Hindus and Muslims.

We reached Dum Dum airport next morning. Although I was now late for the science congress, I had to spend a day in Calcutta on social formalities, including a meal and an overnight stay at Mrs. Banerji's house. Accompanied by an escort, I continued by train to Patna, where we arrived in the late evening of January 3. Waiting on the platform were Banerji, Supreme Court Justice Das, and a few other dignitaries, all dressed in frock coats. Justice and Mrs. Das had been educated in England, and both had traveled widely in Europe. It was a rare pleasure to be their houseguest for the duration of the science congress.

The scientific meetings were not very different from similar gatherings in Europe, though the organization was far less rigorous than at American conventions. There were four especially invited foreigners; each of us was supposed to contribute a popular evening lecture. On the first occasion the speaker was taken to one hall while the audience, quite correctly, foregathered somewhere else in the sprawling city of Patna.

The all-powerful governor of Bihar solved such problems with an ease typical of the great Akbar. The governor's houseguest was no less a person than Sir C.V. Raman, Nobel laureate in physics (1930). He needed little urging to let his voice be heard. Sir C. V. had been awarded the Nobel prize for his discovery of the Raman effect, a peculiar kind of scattering that can best be seen when a beam of monochromatic light passes through a transparent liquid, such as benzene. Sir C.V. explained his work in such simple terms that one wondered why these things had not been discovered by Archimedes. During his visit to Stockholm in 1930, Raman had demonstrated the Raman effect on alcohol. At a gathering the following evening, his Swedish colleagues tried out a similar experiment, but with the process reversed. From the account, I gathered that both experiments had been successful, in that they produced predictable results. Raman had a rare gift of presenting science in a simple and fascinating manner. When my turn came to speak I had a feeling that the audience hoped something would go wrong so that they could hear Sir C.V. again.

From Patna, Banerji and I continued by train to New Delhi where we arrived on Friday, January 9. Again, no real work could be done until all formalities had been cleared. An apartment had been reserved for me at the Congressional Hotel. It was fascinating to observe the activities at this political hub of a multilingual nation whose representatives had to use the English language to communicate with one another. I was pleasantly surprised to meet many politicians who, quite frankly, expressed gratitude for the many democratic institutions and practices that they had inherited from the British. However, it was difficult for me then to measure the depth of the Hindu version of democracy. Up to about five months earlier the British had formed an exclusive top layer; more than ninety percent of the natives were illiterate. In August 1947 the top layer had been replaced by Indian bureaucrats. Nothing much seemed to have happened at the lower levels, however, and the partitioning of the country had created enormous problems for which neither side seemed to be prepared.

Arriving in New Delhi I found that rather elaborate arrangements had been made to fill my time. The itinerary indicated that I would be taken to see the Taj Mahal on Sunday, call on the minister of transport, and sign the visitors' book of the governor general (Lord Louis Mountbatten) on Monday, then call on Nehru in his office on Tuesday, and on Gandhi at Birla House on Thursday. Between these activities I would see the monuments of Delhi, visit the meteorology department and the university, and give a talk on meteorological education.

Delhi, I soon found, was not only a center of politics and ideologies, but also a cauldron of discontent among the underprivileged classes. On Monday (January 12) Gandhi declared a fast to overcome the impotence that he had long felt in his service of the common people. He fasted also to put pressure on the extreme wing of his followers and on the Muslim leaders to force them to collaborate in solving the refugee problem. My visit to Birla House was, therefore, postponed until late February, when I would again be in New Delhi. Although this calling on high officials seemed completely pointless, all I could do was obey the schedule and try not to waste anyone's time. Nevertheless, I was overawed by the thought of meeting Gandhi. I welcomed the postponement, for it would give me time to read and learn more about India.

The visit to the Taj Mahal was made out to be an official function, the pretext being that I was to inspect a hydrogen factory in Agra operated by the meteorological service. The local official in charge took his orders very seriously (most Indians do) and he seemed set on showing me every cock, bolt, and nut in his not-too-modern factory. To catch the last train back to New Delhi, I was left with but little time to see the Taj which, with justice, has been said to be "within more measurable distance of perfection than any other work by man." Where else in the world can one find such breathtaking beauty? I left, wondering which was the more beautiful: Shah Jahan's love of Arjumand, or the Taj, which he had built in her memory.

The call on the minister of transport was much to my liking: short and to the point. Aviation was a central theme, and he was not one of those who considered flying an upper-class luxury. He explained future plans and spoke like a Hindu geopolitician. After less than ten minutes I left with a warm handshake and a number of well-measured phrases. Formalities are there to simplify rather than to complicate matters.

The call on Nehru was quite different: a busy man in a room furnished no more lavishly than an office of a deputy assistant under-secretary in Washington. It was late in the first day of Gandhi's fast, and I felt like a horrible intruder. I had hardly finished my "how-do-you-do" before I began to search for an excuse to leave. After two or three attempts I understood that Nehru did not like to be interrupted. On one occasion, when Banerji tried to explain his meteorological requirements, he was brushed off rather unceremoniously. Nehru wanted to talk about water resources, irrigation schemes, hydroelectricity, fuels, and industrialization. As a young man, before he had found his way to Gandhi, he had visited Norway and had fallen off a ship in the icy waters of the Sognefjord. He was well informed of recent developments of hydroelectric power in Norway, and mentioned plans for giant turbines in the Himalayas. A daughter (or perhaps it was a close relative) was (or had been) studying at Lund University (which he took to be in Norway). Addressing Banerji, he said, "Let us send our meteorology students to Norway."

A silent assistant entered, tiptoed to Nehru's desk, and handed him a scruffy piece of paper. Nehru's face turned ashen. After a considerable pause he burst out, in a mixture of sorrow and anger, "Medical symptoms that turned up on the fourth day of the last fast are here already." And then, incongruously, "What can *I* do?" Never had I felt so sorry for a man as I did on this occasion. For almost three decades viceroys had feared the possibility of Gandhi's dying on their hands. Now, less than five months after the liberation, this same fear had come to Nehru. But there was a difference: Gandhi was now seventy-nine years old, and his frail body had been weakened by recent fasts and marches. "What can *I* do?" Nehru had asked; neither Banerji nor I could utter a word of comfort.

Before I left, Nehru asked if I would speak on their broadcasting system on the scientific and technological problems of India. His nephew, B. K. Nehru, whose office was only three doors from that of the prime minister, would make the necessary arrangements; Banerji would help with information. I worked on and off on the manuscript, and spoke from Calcutta five weeks later.

On January 16 I left for Poona and was comfortably installed in a spacious apartment in the Poona Hotel, which, during the British rule, had formed a center of social activities. Here I found it convenient to change my daily rhythm. A tea-*wallah* would bring me tea and plantains at 5:00 in the morning, and after two hours of writing lecture notes, a water-*wallah* would begin to carry buckets of warm water to fill my tub. On rainy days the bathroom floor got covered by footmarks, and a third *wallah*—one of sufficiently low caste—would come and clean the floor. Breakfast at 8:00 and at 9:00 a staff car would arrive to take me to the Observatory. There was little to do in the evenings except go to bed early, read, or watch a variety of lizards running back and forth on the picture rails in my bedroom.

At the Observatory I was looked after by a servant of unusual stature: tall, handsome, and probably of mixed blood. He spoke English perfectly and seemed to sense what was about to happen. Though his post of duty was outside the door of my third-floor office, he was always present in the yard when I arrived after breakfast or lunch. I soon learned that it would reflect unfavorably on him if I were seen carrying anything, even my camera or briefcase. He had a grandiose manner of saluting and opening doors. I found his many attentions so disturbing that I rarely left my office without compelling reasons.

The audience at my lectures and demonstrations was exceptionally responsive. Though the professional level was somewhat variable, some could rise to great heights. The main obstacle in the way of progress was an exaggerated emphasis on seniority and precedents.

Since I had not been anywhere near India before, I could not claim any local forecasting experience. The success of my mission would clearly depend on my ability to teach and apply general methods. Arrangements were made for me to receive plotted weather charts, which I analyzed according to the so-called Bergen methods; I then prepared my own forecasts. These were presented for discussion and comparison with the official predictions. The map conferences were generally lively; considerations of seniority and precedence seemed to evaporate into an atmosphere of genuine enthusiasm. Contrary to earlier assumptions, the so-called polar front and airmass methods were found to be generally applicable and not limited to the subpolar region where they were first developed. I felt very much at home in the Poona Observatory. Had I been an Indian, Poona would have been my choice.

In response to Gandhi's fast, a peace committee, representing all political communities in Delhi, had signed a pact pledging protection of life, property, and the Muslim minority in the Delhi region. Gandhi, believing that he had won his point, took food on January 18. Two days later a Hindu youth threw a bomb at Gandhi's prayer meeting. The next day Gandhi spoke in defense of the youth, explaining that he might well have thought of Ghandi's tolerance toward Muslims as an opposition to Hinduism. The bomb had exploded harmlessly, the youth seemed confused, and Gandhi did not desire protection; for him the "truth force" was enough. The prayer meetings continued, as before. However, cold snakes were crawling in the grass in many places. On January 30, as I was having dinner in Poona Hotel, the manager, with tears in his eyes, told me that Gandhi had been killed. Shortly after 5:00 P.M., as Gandhi was about to begin his prayer meeting, a Hindu had approached him with signs of obeisance, then pulled his revolver and fired four shots. "He Rama!" (O God!) were Gandhi's last words.

In the evening Nehru broadcast the news to the nation. His words rang out as a pleading echo of the message of the *Katha Upanishad*:

> The sun shines not here, nor the moon and stars,
> These lightnings shine not, much less the fire!

Not only for Nehru, but also for four hundred million Hindus and Muslims, and for all who cared throughout the world, "The light has gone out."

The day after Gandhi's death was declared a public day of mourning. Nehru had advised all who could to go to riverbanks and pray. Though I had been in Poona two weeks, I had had no time to see the city and the cantonment. After an early breakfast I went for a walk. The town looked ominously calm; shop windows and doors were shuttered; here and there shopkeepers sat, silent and motionless, on their doorsteps; the place seemed without life. I felt as if I were walking alone among the waxworks of Madame Tussaud.

Gradually and silently people gathered from houses and side streets, walking about slowly with a general drift toward the river. I had intended to join in the observances on the riverbank but, seeing how the crowd grew denser, I decided to return to the hotel and watch from my verandah.

Suddenly, and for no discernible reason, a wild commotion broke loose. Windows were smashed, shops plundered, houses set on fire, and the packed mass of people surged forward as if they were all moving toward a common goal. In spite of the ugliness of it all, an element of dignity and sorrow was clearly in evidence. Through the grapevine we soon learned that the man who had killed Gandhi was a certain Mr. Nathuram Vinayak Godse, a fanatic and editor of the obscure paper *Hindu Rashtra*, representing a most extreme faction of Hinduism. His press and home were not far from my hotel. The riots, which started about 10:00 A.M., continued until late in the evening. Even thereafter smoldering fires and a state of uncertainty persisted.

For days on end the leading newspapers served as outlets of Gandhi worship, and all persons of consequence spoke about the irreparable harm that the assassination had done to Hinduism and the Indian cause. Few understood that Gandhi, during the last five decades of his life, had undergone a continuous evolution. Under the influence of many great thinkers in the western world, he had developed a philosophy of God and life addressed to humanity as a whole rather than specifically to the Hindu sector. Had the assassin Godse really done *irreparable* harm? What would Jesus have been without the cross? Even in backward India, technology had left its marks: The revolver had replaced the cross, and a cloak of legality was no longer necessary.

In our conversation on January 13, Nehru had repeatedly referred to education. In the context, higher education was of special interest. However, the strength of a true democracy based on the principle of representative government must necessarily rest on the quality of elementary education. On the pretext of teaching the people how to read and write, Lenin, and Stalin after him, taught the children communism. Even Hitler, who did not have the advantage of starting with illiteracy, tried to tinker with education to achieve his wicked aims. In 1948, India had no choice, for Gandhi's thinking was too deeply rooted in the nation, and Nehru and the men close to him were opposed to a politically tinged educational program. In the years that followed, it is Gandhi's India rather than Jinnah's Pakistan that has contributed to democratic progress.

My work in the Observatory came to an end in mid-February. After visits to the universities of Bombay, Madras, and Calcutta, I returned to New Delhi, where a few days were spent in leave-taking. On the last day a clerk brought me a mile-long balance sheet, itemizing my few incomes and many expenses. A *tonga*[90] took me to Connaught Circle where I exchanged the balance for a few memorabilia and yards of Benares silks and gold brocades. Three days later I was back in icy Oslo.

Chapter 25

Full Circle

Half the world wandered through,
Searching where such flowers grew.
—Rudyard Kipling, "Blue Roses"

Even before I went to India, I had been considering a return to the United States. While I was there the problem became reduced to one of diplomacy and timing. Although Parliament had established a position for me, with the almost Göringesque title of *Riksvaervarslingssjef*, (chief of the Norwegian Weather Forecasting Service), and had provided generous responsibility and pay, the administrative arrangements were cumbersome and grew worse as the service expanded. It became clear that, for years to come, most of my time would be consumed in petty administration. I knew that patience was not one of my virtues.

The Dunstable notes, my vade mecums were still gathering dust; my textbook on weather analysis and forecasting, written between 1935 and 1940, was sadly outmoded but, in the absence of a competing text, the book kept selling at an alarming rate. There was nothing I could do to withdraw it from circulation. Clearly, it was my duty to provide an up-to-date revision, but without a laboratory, a small research group, and time, I had to go somewhere else to start.

There were also other considerations. At about this time John von Neumann, with his group at Princeton, was developing the first electronic computer capable of handling embryonic forecasting problems. It seemed reasonably certain that the center of gravity in forecasting research would shift away from Europe, at least temporarily. Since my interests were in clinical rather than theoretical studies, I hoped to find work in the shadowy area between basic research and application—with no borders clearly defined.

Stirrings of a totally different kind were even more powerful. Living in the postwar atmosphere of war-ravaged Europe, I had come to regard the lend–lease agreement, the victory over Hitler, the Marshall Plan, and similar events as political miracles that could hardly fail to dominate future world trends. True, there were still a few dictators at large and western Europe was in a sad state of repair—but the United States had remained strong. While fighting a world war the standard of living there had been raised, and the people, even those of German extraction, had come to realize their responsibility in a rapidly shrinking world. Roosevelt and Truman seemed to have succeeded where Wilson had failed. I had always nourished a desire to be near the center of action. In these circumstances it was easy for me to forget the troughs in American history and to recall the peaks: the indomitable character of Washington, the wisdom of Jefferson, Jackson, and Lincoln, and the skill of Roosevelt. The so-called American dream was very much alive and the United States, a democracy with overwhelming power, would surely lead the world toward higher levels of human dignity. In spite of the appalling poverty I had seen in India, the wave of idealism that had followed in Gandhi's wake would, I thought, add much to the strength of the democratic camp. I could see "democracy on the move," an expanding sense of responsibility by the "haves" toward the "have-nots," and a gradual elimination of sickness and squalor. Who but the United States had both the will and the power to lead?

At the time when I was ready to renew contacts in America, offers from various sources reached me. The most grandiose one was of rather unusual proportions: I was invited to plan, and later direct, an institution for exploration and research in the Arctic—something to match the Soviet effort in this field. Though the institution would be government-supported, it would be privately operated. The pay, opportunities, and prestige would be quite out of the ordinary. Although I was sorely tempted, I decided that this was not for me, philosophically as well as otherwise. No one could enter this field without having to do a lot of political tiptoeing, and I knew that my skill in such maneuverings was totally inadequate. Also, I would have to leave weather forecasting, which hitherto had been my absorbing interest.

As I have mentioned earlier, I had met Don Yates, briefly over the telephone during the Overlord operation and then in person during the Toronto meetings in 1947. Since the war this very young and able colonel had been promoted to general officer's rank and placed in charge of the Air Force Weather Service. Don, with irresistible arguments, persuaded me to join him and take charge of the scientific advancement of his service, which by now had a wider sphere of interest than any other meteorological organization. Don was a technological generalist rather than a conventional meteorologist, an enthusiast with sound judgment, a man of outstanding drive and courage. Only his skill and brilliance had saved him from many awkward situations. He retired early as a lieutenant general.

At the time our collaboration started in late 1948, the air force efforts in scientific research and development were rather disorganized. Don's outspoken interference (sometimes with, and at other times without, my advice) served to promote many useful schemes, some of which were well outside the field of meteorology. One of my categorical conditions for joining Don's staff had been that I would have a small research group as part of my organization. There were, however, formal difficulties. Research was not an authorized activity in the Air Force Weather Service, and the organization chart had to be approved by someone above Don's head. A person well versed in Pentagon practices suggested that I change the word "research" to "evaluation." Thus, without any change of substance, formalities had been satisfied.

Though it might be hard, at any time, to find a more insignificant case of deliberate distortion, the important thing to note is that the practice, which was already widespread, proved capable of growth, and reached gigantic proportions when the United States, under Nixon, fought a "secret" war in Cambodia. However, as early as 1948, relatively small distortions "within the system" were permissible.

My reception in Washington was exceptionally cordial. Soon after my arrival an air force officer invited me to a cocktail party at which I would have the opportunity to meet Henry (Scoop) Jackson, then a young member of the House of Representatives, who was soon to be elected to the Senate. Jackson inquired about my personal problems such as housing, clearance for access to needed information, and other matters that might influence my work; he even offered to have me made a U.S. citizen by Act of Congress. Since such generosity was hard for me to understand, Jackson explained that Bernt Balchen, the Norwegian aviator, and a few others had obtained citizenship that way. Since such matters had neither political nor economic consequences, the House could be expected to pass a resolution unanimously and without discussion. Being a newcomer and having heard and read a great deal about political corruption, I decided to remain at a respectful distance.

In 1949 Congress passed legislation to establish a super grade of civil servants and authorized President Truman to promote 140 of the top government employees. Don and a high-ranking general went to work on my case. The undersecretary of the air force became interested and after his two visits to the White House, I—a newly arrived alien—got in on the president's list. Though my inclusion was out of proportion to the importance of my work, I was, naturally, pleased. So were certain air force officials: They felt that I was now firmly tied to their organization.

In the meantime the overall picture began to change. In 1950 Don Yates was moved to the air force staff where he became involved with the policy direction of scientific activities. A major reorganization was already in the planning stages. The outcome of committee studies was a proposal for the establishment of a major Research and Development Command to be headed by a senior general. The commander would be assisted by four civilian scientists, each of whom would be in charge of a major program area. The four would jointly serve as an advisory council. By early 1952, the plans had matured to the point where Don and others were scouting for talent. For tactical reasons, it was important to "get on the road" without loss of time. At least one early appointment of a civilian scientist would look convincing.

Don had spun his net with care. The commanding general invited me to see him at his headquarters in crowded and untidy Baltimore, a city known even then for political irregularities. The general was well briefed. He knew all about my background and current activities, and a purring coffee machine was there to make me feel welcome. The plans for the new command were explained and so were the enormous benefits that would accrue to science and humankind now that the air force had decided to throw its weight behind scientific research. The general explained that he was especially anxious to have a really progressive effort in geophysics, including the atmosphere up to great heights. His words seemed well chosen. I noted, with satisfaction, that he said "progressive," where other Americans would have said "aggressive." I was not pressed for an immediate answer, but we agreed to meet again soon.

When we met again, the general seemed to take it for granted that I could join his staff within a few days. He had four stars, and my boss had only two, and stars can be very persuasive. Quite frankly, I told this charming gentleman that I had no desire, nor any qualifications, for a job in management. My ambition was to remain in forecasting research. I had already accepted an invitation from the University of Chicago to establish a Weather Forecasting Research Center there, hoping that interested government agencies would contribute to the support of my activity. After a few cups of coffee we agreed that it would be right for me to pursue my own plans. If I should need financial support, he said, the air force would always be interested in my work.

Since the early thirties I had had too many administrative cares. For Chicago I planned only a small group, with hardly any routines. The U.S. Weather Bureau moved their District Forecasting Center into my building so that I would have easy access to all kinds of charts and data. The air force provided a steady financial support of about $90,000 a year, which was all I needed for the activities I had planned. My immediate desire was to concentrate on my Dunstable notes and to conduct such research as would add up to a reasonably coherent ensemble of methods in analysis and forecasting.

My association with the air force remained cordial, and increased in warmth as many of my early students rose to high ranks. After joining the Chicago faculty, I served, on and off, as a consultant. Through this activity I had occasion to visit air force bases in Alaska, the Aleutians, Japan, the Philippines, Hawaii, and eastern Europe. A similar but less demanding association with the army kept me in touch with their research on snow and ice. It also brought me through a snowstorm that lasted for seven days on end at the top of Greenland. On the whole I found these contacts—my liaison with reality—most inspiring.

The University of Chicago, I found, was unique among American learned institutions. Its organizational structure was exceedingly complex, but its basic philosophy was simple and unconventional. Where else could one find such a rich variety of activities in divinity only two hundred yards from a street that had justly acquired its reputation as the "Sin Strip?" Where else could one find a chair in metaphysics, an expert on lava on the moon who showed annoyance when asked a question about lava on the earth, a famous savant who proposed in earnest that the university telephone system be restructured so as to accommodate outgoing calls only, or an absentminded philosopher who kept on typing for three hours, forgetting to put paper in his machine? And where else could one find a major effort with such a utilitarian aim as to improve weather forecasting?

On the sprawling Chicago campus it was easy both to overlook and to discover a number of fringe activities overshadowed by towering excellence in all basic fields—in the humanities no less than in the sciences. And where else had the quality of education received so much attention, all the way from the nursery level to the doctoral degree and beyond?

The university had begun in the early 1890s, not as a school or a small college but as an Oxford-type university. A total of thirty-five million of John D. Rockefeller's dollars had given it a crisp start. The first president, William Rainey Harper, and his successors had, over the years, helped many earnest citizens to part with funds, thus enhancing their usefulness and stature. Most people like to be considerable within their own community, and the university had responded in style. However, as taxation began to climb after the Great Depression and accelerated during the war, the volume of philanthropy tended to decline. To meet the responsibility of greatness, the university began to wear out its presidents at an alarming rate.

In the meantime Chicago and a few sister universities in the east, being free of government control, developed talent, set standards, and helped to raise the level of higher learning throughout the varied United States. Unfortunately, education at the lower levels did not follow suit on a nationwide basis. As material needs accumulated during the Hitler War and technological greed followed in its wake, life became harder for those universities who saw that the by-products of the science of Galileo and Newton had to be balanced by derivatives of the teachings of Socrates and Plato.

While the avenues that went from the science laboratories to the shop windows and military arsenals were wide and numerous, the paths that led from the teaching laboratories to the homes were few and hard to discern. So many impulses came through the mass media from a cultural wilderness of advertising, crime, and vulgarization of sex. Between the steadily expanding domain of the natural sciences and the shrinking area of the humanities lay the developing activities of the social sciences. The sociologists of the postwar years, however, seemed to bend their efforts the easy way, toward the physical sciences and cold statistics. The brutalization process—at home and abroad—resulting from the war received but little attention, and social stresses and strains were building up within the nation. Although Chicago's philosopher-president, Lawrence A. Kimpton, and a few other cultural leaders did not seem to like what they saw, there was little that could be done against worldwide trends in which the United States had ascended to leadership.

I had hardly settled in at the university when I had the good luck to discover a new type of storm development. I say *new*, in the sense that no description had ever appeared in the literature, though, undoubtedly, astute forecasters may well have suspected the existence of an unrecognized type. In fact my Dunstable notes contained references to two strange cases that had developed over the Mediterranean with strong northerly winds across the Alps. Our storm, which struck the Midwest with devastating force on November 25, became widely known as the Thanksgiving Storm of 1952. The facts that it arrived unannounced and did not adhere to orthodox patterns added interest and caused me to take on, as a major project, a comparison between the classical Norwegian scheme and the new type.

In the Bjerknes scheme, to which reference was made in chapter 2, large migratory cyclones develop according to a pattern supported by observation as well as theory. Under certain conditions warm and cold air masses are brought together, creating a difference in potential energy that, in suitable cases, may be converted into kinetic energy (or storm winds), while the temperature contrast weakens. In our Thanksgiving Storm the process was rather the reverse: The temperature contrasts were weak initially and increased as the storm gained in fury. Clearly, here was a mechanism that had not been explored.

My group now put all its effort into research on various aspects of storm development. After much analysis and computation, it was found that the Bjerknes type was the dominant one, except in the general area from the high Rockies toward the East Coast; here, the Chicago type was found to be prevalent. The same may well be true in other regions where major mountain ranges are crossed by strong upper winds.

The atmosphere is a very complex system, however, and mixed cases represent the rule rather than the exception. It was not until late 1967, four years after my retirement, that I was able to identify a pure case, one in which no temperature contrasts were present initially. This case was so spectacular and so tempting that I decided to interrupt my retirement activities. In collaboration with the U.S. Weather Bureau and the National Meteorological Satellite Laboratory, I returned to my storm research. It was easier now, with electronic computers to handle large volumes of data. Although the symptoms of this type of storm development are now well recognized and the location of the energy sources are known in broad outline, no one has been able to develop a theory to cover the case. It is certainly not out of kindness to

Fig. 25.1: Petterssen with students at the University of Chicago.

the younger generation of colleagues that I have left the problem unsolved.

Even after I moved to Chicago, my outmoded textbook kept selling. However, I was now in a position to work over my Dunstable notes, weed out untenable ideas, fit in the results of our storm investigations, and draw on much new material that had been published since the war ended. I could also test my ideas in front of my very bright students and colleagues (Fig. 25.1).

In rapid succession I published revised editions of *Weather Analysis and Forecasting* (1956) and *Introduction to Meteorology* (1958). Without much delay translations of one or the other, or both, appeared in Hindustani, Italian, Japanese, Polish, Russian, and Spanish. The reception was everywhere cordial, and I felt richly rewarded for many years of hard work—a stress that I felt could not continue much longer.

In the summer of 1958, while recuperating from a passing complaint, I received a fine dose of medicine from a most unexpected source. The New York Board of Trade, under their Business Speaks program, had decided to give me their Gold Award, a large and beautifully mounted plaque with an inscription expressing grateful appreciation ". . . for Notable Services in the Preservation of our Heritage of America. . . ."

Fig. 25.2: Petterssen receiving the Gold Award from the New York Board of Trade in 1958.

Since this high honor was completely out of proportion to my accomplishments, the board, I am sure, will forgive me for adding an explanation. The award had been instituted in 1947, and I was a recipient of the twelfth award. In earlier years, outstanding men in law, business, political, and military fields had been honored, including such giants as Bernard Baruch, Winston Churchill, General Eisenhower, Herbert Hoover, General Marshall, Judge Harold Medina, and John D. Rockefeller, Jr. In 1957 the board had honored A. Whitney Griswold, then president of Yale University, for his monumental contributions to education. In 1958 weather forecasting was considered to be an up-and-coming field, and the award, I took it, had been meant as an encouragement to the profession as a whole. This thought was reflected in my acceptance remarks where I made special reference to my many students and collaborators. The award ceremony took place in the Grand Ballroom of the Waldorf Astoria (Fig. 25.2).

As a matter of convenience I had booked a room in this expensive hotel; the cost was $27 a night, fifty cents less than I had recently been paying at a hospital in Chicago. Through the New York Board of Trade I had the pleasure of meeting a large number of the supposedly distasteful capitalists of New York City. I was impressed to find that so

Fig. 25.3: Jack Bjerknes and Erik Palmén receiving the Carl-Gustaf Rossby Award from Sverre Petterssen, president of the American Meteorological Society, in 1958. In the foreground is Mrs. Harriet Rossby, widow of Carl-Gustaf Rossby.

many of them were as thoughtful and individualistic as my university colleagues.

In 1958 I was also honored to be able to *give out* awards. That year Jack Bjerknes and Erik Palmén received the inital Carl-Gustaf Rossby Award from the American Meteorological Society (Fig. 25.3).

During my Chicago years I became increasingly involved with national planning and much of the committee work that surrounded the science anthill in Washington. Thus, after 1950 I was active on advisory committees to the House of Representatives, the Department of Defense, the National Science Foundation, the National Academy of Sciences, and the President's Science Advisory Committee, as well as on a number of ad hoc committees on special problems. The beginning of my involvement in these activities goes back to the time when I was still working for the U.S. Air Force.

Then, rather unexpectedly, weather forecasting and weather modification (often referred to as rainmaking, or even as weather control) became the main ingredients of the same broth, with no shortage of powerful cooks to stir the pot. To appreciate the recipe of this strange dish it will be necessary to recapitulate a rather complex chapter of meteorological history.

In the middle twenties Bergeron had discovered a mechanism that could cause a myriad of minute condensation droplets to join and form an ordinary raindrop. The condensation droplets could do this if they had a nucleus to deposit their water on. Bergeron found that small ice crystals were effective as nuclei. Rain that is released from clouds with ice particles in them is referred to as *cold rain*. In 1939 I had invited attention to other possible mechanisms and published observations from Hawaii showing that on several occasions rain had been released from *warm* clouds. However, the Bergeron effect appealed to meteorological thinking and was rightly considered to be the dominant one in middle and high latitudes, where the clouds often reach up to freezing temperatures.

About 1947 Irving Langmuir—a Nobel prize laureate in chemistry—and his group, working at the General Electric Laboratories, had discovered that silver iodide had a structure similar to small ice crystals. Since natural clouds, even at very low temperatures, are generally deficient in ice crystals, while silver iodide can readily be produced, it seemed possible to supply silver iodide dust to cold clouds, hoping that the clouds might be "fooled into believing" that natural ice crystals were present. If the clouds could be so misled (and few doubted it) weather modification (or control) would not be much of a problem.

Langmuir was unlucky, and became a victim of one of the many pitfalls that nature so generously provides for scientists who venture too far outside their own field of specialization. Though a leading authority on the chemistry of crystal points and surfaces, a philosopher, and a polyhistor[91] in general science, Langmuir did not appreciate the complexity of meteorology as a science. In the atmosphere, processes of vastly different spatial scales and life spans exist together and interact; impulses and energy are shuttled through the whole spectrum of phenomena—all the way from molecular processes to global circulations and the changes in the atmosphere as a whole. No chemist, physicist, or mathematician who has not lived with and learned to understand this peculiar nature of meteorology can pass valid judgment on how the atmosphere will react if one interferes with the details of the natural processes. Moreover, to determine whether or not the atmosphere has responded to outside interference, it is necessary to predict what would have happened had it been left alone.

As I have just said, Langmuir was unlucky. For no profound reason he had left a silver-iodide generator somewhere in New Mexico and made arrangements with a local person to "burn" the generator on a weekly schedule. Using a set of readily available weather reports, Langmuir found that the rainfall had begun to vary in a weekly rhythm. The amazing thing was that the response was not just local; it was nationwide and might well be of hemispheric proportions. Langmuir, and many with him, concluded that the weekly injection of silver iodide from a single generator in New Mexico had excited a hitherto undiscovered natural rhythm of the atmosphere, with the result that the rainfall had yielded to the will of man.

At the time of Langmuir's experiment severe floods developed in the Ohio Valley, causing widespread damage and even loss of life. Langmuir then felt morally compelled to discontinue his exciting experiment; soon the weekly variation in the rainfall came to an end. On one or two occasions before the Ohio disaster, Langmuir had released silver iodide in thundery situations and seen spectacular developments with damaging cloudbursts. In his mind, and in the minds of many others, there was but little doubt that the weather processes could be intensified or repressed to suit human needs.

It so happened that two such needs were immediately apparent. The United States was engaged in a war in Korea, and the Berlin airlift had many awkward weather problems. Though the Ohio floods had deserved special consideration, the theaters of war, hot or cold, were something else. Langmuir took his data to Vannevar Bush, who had served as Roosevelt's science adviser and was now close to Truman. Bush, an engineer and expert on weapon systems, and, like Langmuir, a polyhistor in science, found Langmuir's data very convincing. He wrote to General Marshall, then secretary of defense, suggesting that Langmuir's technique might deserve consideration as a secret weapon. Bush's letter was passed on to General Bradley, then chairman of the Joint Chiefs of Staff. Bradley, who was far more excitable than Marshall, called in his men to explain why this technique had not been properly developed and brought to his notice.

A committee of independent scientists had to be appointed immediately to examine the state of the art and to report without delay. But, reasonably enough, the Joint Chiefs could not talk directly to scientists, nor read their reports; there had to be a "cushion committee" to receive and digest the report, weed out scientific jargon, and reduce the findings and recommendations to a page or two. The cushion committee, which consisted of an admiral, a general, and the chief of the U.S. Weather Bureau, soon agreed on some basic principles, including: One, that no government employee should sit on the scientific committee, and, two, that the fields to be represented were physics, aerodynamics, statistics, and meteorology. However, they were unable to agree on the choice of a chairman, except by violating the first of the overriding principles. Although I was employed by the air force, I was appointed chairman of the committee, which was to be known as the Ad Hoc Committee on Artificial Cloud Nucleation—a name that did not suggest interest in secret weapons. To add camouflage, Dr. A. T. Waterman, director of the National Science Foundation, was appointed a member.

The year was 1951 and the administrative rumblings of the atom bombs dropped on Japan were still reverberating in Washington. Everyone knew that Enrico Fermi was a famous nuclear physicist, and, in a guarded telephone conversation, where no details could be discussed, a member of the cushion committee managed to persuade Fermi to serve on my committee. A few days later, when Fermi had discovered that our interests were centered on *ice nuclei* rather than *atomic nuclei*, he begged off. I was allowed to assist in the selection of a replacement, a less famous but exceedingly useful team worker. The committee visited every research group that had professed an interest in the physics of clouds. All, including Langmuir and his enthusiastic disciples, agreed that no technique really existed, though they foresaw considerable results for their further researches. Clearly, the Korean War and the Berlin airlift would have to continue along conventional lines.

The cushion committee next instructed us to design and supervise a program of experiments and tests to determine the potential value of artificial cloud nucleation (cloud seeding, for short). Again consulting all interested groups, a set of agreed programs emerged. As expected, the documentation of the results of experiments on natural phenomena raised intricate problems. Some of the experimenters soon found that they had become reduced to meteorological lambs among statistical wolves. The committee, however, remained unyielding. Statistical controls soon became well accepted, and the field experiments progressed smoothly until the funds began to dwindle.

The results of our labors proved inconclusive, in the sense that neither positive nor negative effects had been detected with any degree of certainty. It was clear, however, that our understanding of the basic processes associated with condensation and precipitation was so meager that extensive research had to be carried out before attempts at weather modification could be justified. As to the Langmuir periodicity, our verdict was that, once in a while, the atmosphere over limited regions may behave in a manner that in some respects appears to be repetitious. On many occasions in the past, so-called periods had been "discovered," but almost every one of them had disappeared after a while, leaving behind a disappointed searcher for some simple truth of nature. Langmuir had been especially unlucky, for the data he had used showed a clear periodicity, neatly tuned to the rhythm of his silver iodide injection. While on this occasion Langmuir had certainly misinterpreted his observations, his name as a giant in science has remained untarnished, and the inspiration that he so generously spread around has made it possible for many lesser men to be remembered.

Although the results of our field experiments were inconclusive, the Langmuir controversy lived on and stimulated new endeavors in basic research in Australia, Europe, the Soviet Union, and elsewhere. Undoubtedly, the results of these activities will soon make their influence felt in weather forecasting. They might, in the fullness of time, lead to some degree of modification of important weather phenomena. How much and how soon, no one can say. Clearly, interference with natural weather phenomena, even if it can be achieved, will serve no useful purpose unless it can be performed in a predictable manner, in response to a real human need, and without causing unacceptable damage to other interests. Costs, too, have to be considered.

Nevertheless, impatient minds kept pressing for shortcuts to beneficial results. Some consultant firms offered rainmaking services to utility companies and farmers in rain-deficient regions, and politicians began to press Congress for legislation and financial support for weather modification activities. Soon, the confusion increased beyond reasonable limits. In the late fifties and early sixties, some organizations were engaged in the prediction of unmodified weather, while others offered to modify unpredicted weather. However, after two decades of work and an expenditure of a few hundred million dollars, the practical results of the shortcut approach to large-scale weather modification seem very much in doubt. Although the research is now progressing in a more orderly fashion, the pressures of shortcut approaches and commercial exploitation were very strong at the time when I became heavily involved in science planning.

On October 4, 1957, the Soviet Union launched the world's first space ship, *Sputnik I*. Soon, a kind of panic spread through the United States. The fear had little to do with real human needs at home or abroad. It was not concerned with poverty, health, environment, education, or the forward march to higher levels of human dignity. Rather, it was rooted in the competitive attitudes of the two superpowers and the desire by each for undisputed dominance in a divided world. Soviet scientists, working under strict state control, had excelled in a narrow field—brute propulsion. Although such propulsion was far beyond real human needs, the horror aspects, the military applications, could not be ignored. *Sputnik I*, an almost empty uninstrumented shell sending out conventional radio bleeps from its orbit in near space, triggered far-reaching responses. For political and military reasons and for the prestige value of it all, the American effort in science and technology had to be doubled and redoubled. An entirely new hardware industry—the space industry—had to be created. Fast-growing mushrooms were seen to spring up overnight, and the Washington anthill grew beyond recognition. Rarely, if ever, has it been so difficult to distinguish between real and imaginary values.

Immediately after the launch of *Sputnik*, Eisenhower appointed James R. Killian, president of MIT, as Special Assistant to the President for Science and Technology (SAPST); he named a top committee, the President's Science Advisory Committee (PSAC), to guide the national effort. There followed a period of national promotion and indiscriminate support for science and technology, with heavy emphasis on fields related to space and propulsion. Although this one-sided expansion and funneling of young talent into prestige programs resulted in enormous advances in the preferred areas, the very cultural process suffered, partly through neglect of the humanities and partly by overspecialization of science teaching. These were the inevitable results of the hasty response to *Sputnik I*, following so closely after the dislocations caused by World War II. Again, *more*, *faster*, *money,* and *manpower* had become the buzzwords.

In 1959 George B. Kistiakowsky, a Russian-born Harvard chemist of world fame, succeeded Killian as SAPST and chairman of PSAC. Although most people tend to admire bigness and prefer optimistic messages, Kistiakowsky appeared to nourish doubt about the wisdom of indiscriminate expansion. He taught us to grade our adjectives, with "good" at the lower end of the scale and "excellent" reserved for that which is preeminent. As the saying goes, all makes of beer may be said to be good, but few are really excellent.

In the United States, science and technology have always been considered as an "all-man's land," with the research effort being scattered through almost all government departments. Kistiakowsky, who had an eye for innovation, instituted a new procedure by appointing panels to review and research requirements, efforts, and expenditures for each field rather than for each department. In this manner duplication of effort and poorly justified activities, as well as high quality work, could be more readily identified. This reorientation soon led to long-term planning and outlines of national goals in scientific research. These activities became intensified when President Kennedy moved into the White House and Jerome B. Wiesner, director of MIT's Research Laboratory of Electronics, succeeded Kistiakowsky.

The reviews initiated by Kistiakowsky and expanded by Wiesner led to studies of long-term goals in some of the less tidy branches of the sciences and to estimates of human resources to support the research. Although meteorology was recognized as a much-neglected area, its international flavor appealed strongly to President Kennedy, particularly because he was anxious to seek out new noncontroversial fields in which the United States and the Soviet Union could agree on joint efforts for the benefit of the world as a whole. To Kennedy, the term "superpower" meant more than just "superforce," and his thoughts ranged beyond the narrow confines of tolerated coexistence.

In June 1961 Weisner arranged with the National Academy of Sciences that I be assigned to direct a major planning effort, the aim of which was to develop ten-year plans (1961–1971) for what henceforth was to be called, not "meteorology" but "the atmospheric sciences." The change of name was meant to indicate an upgrading by Washington of what Aristotle had called the sublime science.

The job before me was obviously a difficult one. I was fortunate enough to secure the services of Colonel Marshall Jamison as my assistant. I knew him well; he had been my executive officer while I worked for the U.S. Air Force. I had often relied on him for hard work and sound advice. Within a week, invitations went out to close to 200 leading scientists in meteorology and related fields. Experts, some of whom came from Canada, India, Israel, Mexico, and several European countries, gathered at six consecutive planning conferences.

By the end of August a heavily documented report in two large volumes was ready for processing by the Academy and for transmittal to Wiesner. Although the theme of the report was national goals, an important section dealt with international aspects, the reason being that so many of the problems in meteorology are of a global nature.

In the international sphere the report proposed three major collaborative efforts:

1. An *International Atmospheric Science Program* (IASP) to further research on problems of global or multinational scale.
2. An *International Meteorological Service Program* (IMSP) to assist the less-developed countries in providing the meteorological services needed for their economic and social advancement.
3. *A World Weather Watch* (WWW) to provide, on a highly flexible basis, the observations and telecommunication services needed for the science and services of meteorology.

The report stressed the need for active collaboration among several of the specialized agencies of the United Nations, especially the WMO and UNESCO, the U.N. Educational, Scientific, and Cultural Organization.

In the meantime, while the report was being processed by the Academy, other things happened. As President Kennedy had begun to prepare an address, his first, to the General Assembly of the United Nations, Wiesner asked me to collect a small ad hoc committee and prepare a brief report from which could be distilled material for the president's speech. A committee of six met in Washington on September 1, 1961 and prepared a neat document that, in essence, absorbed the three recommendations just mentioned. Pressures from the so-called weather modification lobby were quite strong, but the committee would not go farther than to recommend research to determine the extent to which natural weather phenomena may be modified by artificial means.

The next day, in the relaxed Saturday atmosphere of the Department of State, I briefed a group of U.N. specialists on the report and helped them prepare a précis that could go to the president. It soon became clear that the diplomats were looking for some "wonderfulness" to match the stature of our beloved president. A world center for weather forecasting (situated in Washington) and a major effort in weather modification were among the ideas afloat. My view was that our president would be better served by being provided with a plan for acquisition of new knowledge rather than a set of ready-made and over-advertised technologies. My graceful and very able companions soon agreed, and the précis reflected loyally the substance of the committee's report.

President Kennedy addressed the U.N. General Assembly on September 25. In the clear context of things to come, he announced: ". . . We shall propose further cooperative efforts between all nations in weather prediction and eventually weather control. We propose, finally, a global system of communication satellites linking the whole world in telegraph, radio and television."

The United Nations lost no time in responding. In December 1961 the General Assembly passed its resolution 1721 (XVI) on outer space which, in part C:

> *Recommends* to all Member States and to the World Meteorological Organization and other appropriate specialized agencies the early and comprehensive study, in the light of developments in outer space, of measures:
> (a) To advance the state of atmospheric science and technology so as to provide greater knowledge of basic physical forces affecting climate and the possibility of large-scale weather modification;
> (b) To develop existing weather forecasting capabilities and help Member States make effective use of such capabilities through regional meteorological centers.

The reference in paragraph (a) to weather modification reflects some success by these enthusiasts in influencing U.N. ambassadors to believe that not only the weather but also the climate could be made to yield to the will of man. Although many did not share this optimistic view, no scientist could reasonably maintain that the possibility of such modification should not be appraised, particularly since the potential value of any degree of useful modification might be enormous.

In response to the U.N. resolution, the National Academy appointed a committee, with me as chairman, to formulate specific programs that American officials could take to the international conference tables. On the face of it, my assignment seemed a simple one. The three major programs—IASP, IMSP, and WWW—would obviously form part of any general scheme, but the key to success depended on the availability of trained personnel.

The fact was that in the United States and the other highly developed countries, there was an acute shortage of science students. Those that were available were more readily attracted to the glamour areas of space, propulsion, atomic energy, and electronics.

During my visit to India in 1948, and on a similar tour of South America in 1960, I had been dismayed to see the vast untapped source of talent that, because of totally inadequate educational systems, seemed doomed to live in ignorance and dissatisfaction—masses of people who could, perhaps, cause political unrest. Scientists, who are past masters at generalizing, will readily agree that "first things come first." Nevertheless, they often disagree when it comes to assigning practical priorities. Although the detrimental effects of inadequate educational systems in the underdeveloped regions have long been recognized, the advantages to multinational industrial concerns of maintaining large sources of cheap labor have also carried some weight. Less appreciated has been the deep-rooted hunger for higher education that has emerged in some of the developing nations since industrialization began to make its influence felt. Not only did this hunger exist among young potential leaders, it had also become fashionable to express it in terms critical of the attitude of certain "have-much" nations. In South America the juggernaut multinational concerns were sometimes thought of as potential instruments of political interference, even to the extent that the sincerity of the foreign aid program was questioned.

In Washington, it was easy enough to find learned reports on how the basic sciences beget application and progress—eventually. In the underdeveloped regions, I thought, it would be far more productive to place initial emphasis on the sciences that are concerned with the human environment and the resources of our planet. My argument was that a concerted effort in these areas would immediately promote the less favored people to essential members of the world community of nations. Furthermore, such efforts would pay early dividends through improved agriculture, fisheries, transport, and commerce. From such beginnings, popular and political support would soon follow and a self-developing process would soon spread to other areas of education, science, and commerce.

However, since my assigned area was international programs in meteorology, I found it entirely within the spirit of the U.N. resolution to recommend, in addition to IASP, IMSP, and WWW, a fourth major component: an International Meteorological Educational Program (IMEP), with the aim of training meteorological scientists and technicians for the less developed regions. It was my hope that similar actions would follow in other branches of the environmental and resources sciences. Meteorology, I thought, was in a good position to lead the way toward a real one-world program.

My direct association with these international programs ended with my retirement in 1963. It was not for me to accompany my proposals to international conference tables. With a view to my earlier interests and international work, I was inclined to consider the World Weather Watch as an extension of the ocean weather station program that I had helped develop immediately after the war. Equally important, in my view, was the educational program. For the world as a whole, real progress would surely depend on adequate data and new talent.

In the decade that followed (1963–1973) great progress has been made in certain areas, much of it due to the indefatigable WMO. I was honored to receive their gold medal and award in 1965 (Fig. 25.4). Although support for WWW by member states has been varied, the observation networks over oceans and sparsely populated areas have continued to expand, and the value of the information provided by meteorological satellites has exceeded our initial expectations. Few will now say that the ocean weather station program and the post-*Sputnik* expansions have not paid handsome dividends, in terms of new knowledge as well as general services, disaster warnings, and other desiderata.

Much success has attended some of the basic activities that would have formed part of IASP, had that designation been retained. With much foresight and imagination, leading scientists have planned and implemented a major activity known as GARP, the Global Atmospheric Research Program. This international program aims to study in depth the grand currents of the atmosphere and their responses to solar radiation and interactions at the surface of the earth. Although GARP is designed to meet the aspirations of a relatively small group of leading scientists in the advanced countries, one must hope that, eventually, the new knowledge will spread to less favored regions. It is to GARP and its activities that we must look for basic improvements in weather prediction, particularly for longer time periods.

Less success has been achieved in stimulating widespread interest in the other sections of the original research program of IASP, such as atmospheric chemistry, air pollution, and biological responses to a variety of atmospheric changes. Dirty air (and water) had received but little

Fig. 25.4: Petterssen receiving the World Meteorological Organization's gold medal and award in 1965 in the Parliament building in Stockholm. On the right, WMO director and president Dr. Alf Nyberg.

attention in our academies until the environmental crisis was upon us. Even thereafter the universality of the problem has not been well recognized.

Least success has attended my proposal for an educational program (IMEP). Although some of the universities in advanced countries have willingly opened their doors to students from developing regions, they have also eagerly offered employment to the most promising graduates to alleviate the personnel shortage in their own ever-expanding and technologically intensive programs. Clearly, any educational scheme for the benefit of developing regions will be doomed to failure unless the students can return to satisfying and productive activities in their homelands. The WWW and the IMSP could have become highly effective in providing the employment opportunities that are essential for the success of a one-world program.

In the early sixties, while I was working on the ten-year plans and the associated international programs, some well-meaning helpers sent me a number of "inspirationals"—exalted pronouncements on science and scientists that might serve as a garb of elegance for my reports. Although an anthology of such statements might be of considerable interest, a few examples will suffice to illustrate a peculiar type of schizophrenia that appears to have been deeply rooted in the scientists of all ages. First, a passage from the Parliament of Science in 1958 (my italics):

> . . . science is entering a new and accelerated state of advancement which will give *man* the possibility of control of his *environment*, over himself, and over his *destiny*, which we have as yet only vaguely sensed.

Second, from a paper before the Communist Party in 1960:

> Under the Socialist system of economy, scientific and technical progress enables *man* to employ the riches and forces of nature most efficiently in the interest of *the people*. . . . to develop the means of *weather control* and to master outer space.

Third, Arnold Toynbee, with his unequalled authority on comparative studies of human developments, found that, in the distant future, our century will be remembered, not for its bloody wars, for splitting the atom, or even for the reduction of disease, but for having been the first age since the dawn of civilization when *man* first thought it practicable to bring *all mankind* the *benefits of his civilization*.

Of the many offerings that I received, there was only one that I felt inclined to use. Being a verse from a modern hymn [*A World Transfigured*], it was obviously not meant for the marketplaces of science and technology:

> They shall rule with wing'd freedom
> Worlds of health and *human good*;
> Worlds of commerce, worlds of *science*,
> All made one and *understood*.

Clearly, what one feels inclined to sing in places of worship, be they cathedrals or simple country churches, is not meant as earnest projections into the unknown, nor as finery for banquet speeches.

In our century there has been an increasing tendency to overestimate the importance of the numerical sciences and to assign to them a monopoly of understanding. The numerical sciences produce many things, including the merchandise for our shop windows and military arsenals. But real understanding, and what we may loosely call the "human good," comes from a much wider field. In certain areas numeration may not even be important. Clearly, human progress cannot be measured in terms of the achievements in the numerical sciences. To see things as they really are, we must learn to separate a few grains of gold from a large volume of dross. Although the grains of gold may be few and far between, they are, nevertheless, the only substance on which we can safely build.

As to *man* and *environment*, as referred to above, it is useful to recall that Langmuir, Vannevar Bush, and others offered weather control as a tool of war, a blind weapon to be used indiscriminately over regions of undetermined size. If we are to believe recent reports from Washington (which may not be justified), rainmaking operations were actually employed in Vietnam and, presumably, also in Nixon's secret war in Cambodia.[92] If we, as Toynbee must have done, look backward over recorded history, we shall find that scientists—the overwhelming majority of them—have always and for various reasons been the eager handmaidens of war and destruction, that wars have caused science to prosper, and that many have tried to justify past wars by the technological advances they brought with them.

We know that Alexander the Great had a group of philosophers to advise him on how to fight battles more efficiently; even at that early date, psychological tricks were not unknown. Archimedes, a pure mathematician with a dislike for applied science, advised Hiero on how to fight the Romans. On a lower level, Marco Polo, a leading engineer and discoverer of intricate laws in physics, sold his services to Kublai Khan. Leonardo da Vinci and Michelangelo, both renowned innovators, were pleased to serve their princes as directors of great war projects, as did Goethe centuries later. The great Galileo sold his telescope and many other contrivances with arguments that have a remarkable resemblance to those used by Pyke in the briefing of Churchill and his companions [chapter 12]. Napoleon, even before he ascended to the summit, collected a group of leading archeologists, not to advise on the fighting of battles (only Wellington could do that) but to explore the most effective way of robbing Egypt of art treasures. Within the living memory of many, Einstein (and others) urged Roosevelt to authorize the development of the atom bomb, and Fermi brought the project to fruition.

It is of some interest to note that scientific activities have developed so rapidly since about 1940 that close to ninety percent of the scientists who have ever lived are still living. Toynbee and other optimists might be right if the behavior pattern of our generation of scientists had shown even the slightest tendency toward more humanitarian aims. Unfortunately, a drift in the opposite direction appears to be much in evidence, at least in the physical sciences.

In the final analysis, the decisions are made where the money is. Few scientists or university presidents will bite the hand that feeds them. In most leading countries the defense establishments have been the most prolific supporters of research, and have thus acquired a position of intellectual overlordship. Most scientists are absorbed in a specialism, each in his narrow sector of learning. Their ambitions are centered on pay, personal dignity, national prestige, and international fame. They are neither for nor against war; they are fundamentally indifferent. As a rule, they are loyal to the government they serve. We need poisonous gas, and scientists will produce it; we need jamming devices to mislead the enemy, and scientists will produce them; we need atom bombs, for we cannot afford to lose the race with the enemy, and scientists of gigantic stature will find the solution and produce them.

There is no Hippocratic promise or oath to guide or hinder the scientist in his choice. When Newton proposed that scientists be sent to the battlefields to learn how to enhance their usefulness, he was certainly not pretending to be a Socratic; his advice reflected his concern for British interests. When Claus Fuchs betrayed the British and provided information on the hydrogen bomb, he was motivated by loyalty to the communist ideology, even to the extent that he bit the hand that offered food and fame. Throughout the history of science we find that loyalty to a chief, a prince, or a government, rather than devotion to the "human good," has been an overriding consideration—loyalty, limited in scope far less than in quality.

There are exceptions, however, perhaps more numerous than those that are bared to our knowledge. When Einstein urged Roosevelt to authorize the development of the atomic bomb, he was motivated, not by loyalty to one government against another, nor by hatred of a sadistic persecutor of political enemies, but by loyalty to the "human good"—to save humankind from the horrors of a Hitler victory. The tragedy was that there were no strings attached; the decision to use the bomb on Japan was a political decision, unhampered by knowledge and considerations of relevance, and unmindful of the "human good" and the echoes of history. For Truman, the narrow basis of national loyalty and instant success sufficed.

But old trees bow down their heads as new winds are passing by. With the emergence, after the Hitler War, of two superpowers, new problems have come to the fore, some with great clarity. The term "superpower" can have but little meaning unless we also recognize the existence of "subpowers," nations with limited sovereignty. In the Soviet sphere the principle of limited sovereignty has been openly declared, enforced, and kept alive in case of new needs. There is no deception here: The words mean what they say.

On the American side things are different. The principle is undeclared, manipulated through multinational juggernaut industrial concerns and a variety of channels and government agencies, and the application is not limited to the Atlantic alliance. Though the contours are far from clear, the Europeans, unaware of the Watergate scandals and the White House horrors of crime and corruption, are supposed to believe that the president of the United States, sitting in the Oval Office, is sufficiently well equipped and informed to press the atomic button on their behalf.

Before our century draws to an end, China, if she so chooses, will attain superpower status and rule over a third group of nations with limited sovereignty. Both China and the Soviet Union are opposed on principle to democracy. There is no deception about it; political hypocrisy is not among their major faults. Far more complex, and indeed dangerous, is a political system that operates under the disguise of democratic practices, a system where elections are manipulated, favors traded, and extralegal practices tolerated, while the effects on social structures and respect for honesty are passed on to coming administrations and generations.

For America 1973 was a sad year, a year of crises and extremes. But, again, we cannot extrapolate from a peak or a trough; trends are trends, and aberrations, large and small, are superimposed on them. The monuments we build are rarely representative of average states. Jefferson formulated his political philosophy and promoted his doctrine to an axiom: "We hold these truths to be self-evident." Although we may return to Jefferson for wisdom and strength, especially when things go wrong in the White House, his doctrine does not lend itself to extrapolation of events in a changing world. Likewise, though we may reap beauty and comfort from Lincoln's sermon on the battlefield of Gettysburg, we find new pointers toward future trends. Each administration has to be pruned of its anomalous growths.

However, as we descend from the plateaus of Jefferson and Lincoln and approach more common levels, the signposts become clearer and, indeed, more practical. As late as 1930, when the Great Depression raised problems in all human departments, Henry Ford declared, "We now know that what is economically right is also morally right." In spite of customary political hypocrisies, the Ford doctrine has provided steady guidance and occasional extensions. Akin to Ford's doctrine is Haldeman's promise: "I shall approve anything that works." And in line with this policy we may note Ehrlichman's assertion that the president of the United States has an intrinsic right to order the burgling of a doctor's office to obtain information on a "political enemy."[93] Descending still further, with emphasis on human rights, we come to the now-famous Nixon doctrine: "You have a right to know: I am not a crook."

Although the early seventies were years of political depravity in the United States and confusion throughout the world, there were also signs of hope. A new breeze began to blow and a few strange trees kept their heads high. On the world stage appeared with greater force two towering figures: Andrei Sakharov (an atomic physicist) and Alexander Solzhenitsyn (a philosopher-writer), both men of courage and principle and both opposed to tyranny. As we leave 1973 behind, it seems possible to believe that there will emerge some kind of fusion, not between the physical sciences and the humanities (which should remain separate fields) but between the philosophies that underlie both: a real concern for the "human good." Will Sakharov and Solzhenitsyn succeed? Will their followers succeed when the masters are gone? Will the breeze from Moscow begin to turn anemometers in other lands? Will the scientists of the world continue to be satisfied with a substate–superstate system, or will they develop loyalties on a higher level—a level where the numerical and the nonnumerical sciences naturally merge?

Early in my life I developed a habit of seeking out a sanctuary for myself—a place where I could be alone with my problems. My first sanctuary was a huge rock on the western shore of a lake about a mile from our little farm at Vik. Geographers would call it an erratic boulder, for it must have been carried there by distant glaciers and left behind when the ice melted. On sunny days the rock was warm and pleasant, and it always provided shelter against driving rain. It was at this rock that I discarded my childish ideas of the devil and hell, and it was there that ideas formed that later developed into a personal god: something wise and powerless, something embodying what we have referred to above as "the human good."

After our retirement we found a small house on the fringe of Richmond Park, Surrey, with a fine view over meadows and trees.[94] Soon the birds, the squirrels, and the deer became our friends, and visits to the park became frequent. Tucked away in the eastern corner of the park is the Isabella Plantation, a spring garden with masses of heaths and heathers, azaleas and rhododendrons, camellias, and a variety of flowering shrubs. Some overfed ducks occupy the pond and make excursions among the flowering water plants in the brook—a rare place for peace and thought.

For several years I had no problem to take to my new sanctuary, but with my pen and a pad of paper, and crumbs and broken biscuits for the birds, I could always justify visits. Gradually the Vietnam War became a problem. I tried to "explain it away" as an extension of the chain of events that began with the lend–lease arrangement and continued with the Marshall Plan, the Truman Doctrine, the involvement in Korea, and so on. If Vietnam was wrong, what was wrong in the chain of events that led to it? Then came rumors, and later news of the My Lai massacre, the fake reporting by the military commanders, the connivance in Washington, the awarding of ribbons and medals for plain murder, and so on. Justice was no longer blind. It could clearly see the differences between corporals and generals, and between generals and political officials. The Pentagon had become a superpower within the system, and only congressional committees that handle funds had real power. Most frightening of all was the lukewarm response by Congress and the nation. In rapid succession came the 1971 election campaign, the mysterious elimination of Senator Edmund S. Muskie as a leading candidate, the Watergate burglary, and Nixon's landslide victory. The fact that corruption and political sabotage—and worse—had been perpetrated on a gigantic scale was clear, but, with the exception of some members of the younger generation, mainstream public response remained indifferent. The nation seemed to have absorbed Haldeman's philosophy and responded with a resounding confirmation: We shall approve anything that works.

In early 1973 my visits to the Isabella Plantation became frequent. While the bushes and shrubs spread their beauty I hardened to a decision, a personal decision, to withdraw. At the age of seventy-five, and with a heart that worked with only fifty percent efficiency, all I could do was use my pen:

Letter to President Nixon June 5, 1973

Sir:

I am driven by shame to resign my citizenship of the United States of America, effective at the end of December 1973. If you have resigned, or been removed from office, by that date, my resignation will be withdrawn.

In my opinion the administration that you have commanded has been basically corrupt, and I consider the same to be true of a political committee with which you have been affiliated. Worse still, the White House, under your command, has been one of the major centers of organized crime—crime against the democratic processes, the democratic institutions, and against our two-party system—in fact crime against what I believe to be the pillars on which human dignity, as a political ideal, rests. Under your administration and leadership, politics has become an advertising and image-selling process, and the respect for honesty has been immensely lowered, with results that are glaringly clear.

I went on to remind Mr. Nixon that Herr Hitler, too, had been elected to office, and that, as far as is known, he never personally violated any German law. Nevertheless, his moral responsibility was enormous. I also reminded Mr. Nixon that Hitler had succeeded because the cultural leaders—in religion, in the humanities, and in the sciences—had not had the courage to oppose him. Still, believing in the resilience of the American people—in the silent majority—I ventured to predict that public opinion and, eventually, history would pass judgment on the Nixon regime. Jefferson and the other founders, I thought, had left behind imperishable values.

As the bells of London chimed in the new year of 1974, I was content to have become a conscientious objector in a nonmilitary sense, a stateless person, free of compromise and without allegiance to a superpower where human dignity has ceased to be a political ideal. Although spring was still far off, I began to pay frequent visits to the Isabella Plantation, where I had drafted the letter to President Nixon. On one occasion, when I was walking back to the car park, I managed to get some grains of sand into one of my shoes. I sat down on one of the benches outside the gate, and as I shook out the sand, I saw that the sole was worn through. Without remorse, or even regret, I whispered to myself: “The

Fig. 25.5: Petterssen lecturing at MIT.

first hole in my last pair of shoes." However, spring progressed with biological accuracy and the plantation dressed itself in finery that surpassed anything I had seen before. With Harriet Löwenhjelm,[95] I could ask, "Is not all quite good? Has not God created lavishly? And has He not gone to great trouble about it?"

Appendix A: Petterssen's Correspondence with J.M. Stagg

Dr. Petterssen to Dr. Stagg March 27, 1972

I have with great interest read your recently published book Forecast for Overlord, and in particular noticed the statement that starts on line 12 from the bottom on page 19 and ends with line 6 from the top of page 20.

I would like you to inform me of the background, in documents from the Meteorological Office or any other material, for associating me, and also indirectly my colleagues in the Upper-Air Branch, with the generation of and/or issuing of forecasts "for two, three or more days." A connection of this nature also seems to be suggested in your conference of January 1944 with Professor Rossby and his colleagues, as expressed on pages 23 and 24 in your book.

I assume that the above-mentioned assertions are somewhat based on actual occurrences. Seeing that, in the context in which they appear in your book, these assertions shed an unfavorable light on my professional competence and judgement, I hope that you do not find it unreasonable that I ask you to share with me the actual facts your assertions are based on.

I would also appreciate knowing if the quotation marks used in your book to indicate parts of my contributions to your conferences are meant to indicate true reproductions of my statements.

Dr. Stagg to Dr. Petterssen March 30, 1972

Your letter dated March 27 has been forwarded to me from Shepperton. I regret that you dislike my references to your work as the leader of the Upper-Air Branch (or Group) in Dunstable during 1943/4 as they are represented in my statements on the top of page 20 and on page 24 in my book. You characterize my innocent (and, as I had intended them to be, laudatory) references as "assertions." Assertions of what, or against whom, for goodness sake?

There cannot be any doubt about the fact that you were the leader of the U/A Branch. Likewise it is inarguable that one of the group's functions (if not its primary) at the time in question was to assist Dunstable in its role vis-à-vis the other forecasting centers, in particular the American center, by developing forecasts for the Supreme Commander of the Allied Forces. At the time when the group was established, Dunstable did not have opportunities for development of forecasts. I am therefore hardly guilty of making unreasonable assumptions when I, with a popular book in mind as opposed to a scientific report, described the group's tasks as "to explore and utilize the implementation of new upper-air knowledge in forecasting for two, three or more days." Nowhere in the book have I, as far as I can remember, made accusations or assertions against your leadership of the group and the value its work brought about.

I must therefore admit that I am completely without an understanding of your reaction to the phrasing you quote. I have read your letter repeatedly in order to find an explanation. In the second paragraph you ask for "the background for associating 'you' with the Upper-Air Branch . . . with the generation . . . of forecasts for two, three or more days." And in the next sentence "a connection of this nature also seems to be suggested in your conference of January 1944 with Professor Rossby. . . ." The way you correlate and repeat the thought of your association with U/A in this paragraph makes me draw the conclusion (since I otherwise do not know where you are going) that your objection must be a result of my associating your name with the U/A Branch. If this is the point in question, I apologize. I had no idea that you wished to be disassociated with the work of the Group/Branch.

In regard to your last paragraph I am certain that you are familiar with the significance of the use of quotation marks in books of this kind, in particular when the preface informs that the text is based on excerpts from diaries and notes.

Dr. Petterssen to Dr. Stagg April 12, 1972

I thank you for your letter of March 30.

I regret to hear that you are of the impression that I wish to be disassociated with the Upper-Air Branch within the Meteorological Office, and from the work of the Group/Branch during the 1941–1944 period. Quite the contrary: I am proud of the work performed by this Group, and am very grateful to the very skilled men who helped in this work. My letter was brought about by a wish to avoid misunderstandings, and the disputed issues can be reduced to a very simple point: What is the background, in documents from the Meteorological Office or other material, for the following statement on page 20 in your book?

> Dr. S. Petterssen, a Norwegian meteorologist who had recently held an academic position in the U.S.A., was invited into the British service to take charge of a unit at Dunstable whose job was to explore and exploit the use

of this new upper air information in making forecasts for two, three or more days ahead.

To simplify matters even further, let me make it clear that the point in question is not what you imagined my tasks to be, but the actual circumstances leading to my invitation. I am certain that you must be as interested as I am in solving misunderstandings. If you would be so kind as to give me a direct answer to the above-mentioned question, other misunderstandings might be solved without larger difficulties. I am of the opinion that you should be able to give me a simple answer to my question, and I cannot understand why it should be necessary to distinguish between what is suitable (as you say) "with a popular book in mind, as opposed to a scientific report."

I am grateful for your explanation of the significance of your use of quotation marks.

Dr. Stagg to Dr. Petterssen April 17, 1972

I wish you would believe me when I say that I would also like to solve and remove any misunderstandings. I do not now have, nor have I earlier had, a wish to present a faulty representation of the significance and value of either your personal contributions to the forecasting conference, or of the contribution the Dunstable Upper Air Branch made through you or Mr. Douglas. I am still uncertain as to what exactly the nature of the misunderstanding on your side is.

You say "that the point in question is . . . the actual circumstances leading to my invitation." I of course do not know the circumstances leading to your invitation. There has never existed any sensible reason for me to know this. All I knew, and all I now know, is that you became a member of the staff in Dunstable during the war, and that you therefore must have been invited. Nobody could have forced you. Secondly, I knew that Dunstable at this point was strongly encouraged to take part in the conferences that led to the generation of forecasts for longer time periods for SHAEF and other military purposes. Soon after you became a member of the staff, you chaired a new group established to aid this work by making use of the new upper-air knowledge. I therefore assumed that you were invited to join the staff for this purpose.

If this assumption is wrong, and if the phrasing of this paragraph does you wrong in a serious way, I apologize, though I still do not understand how I am doing you wrong.

All I can say and repeat is that I sincerely believed that the sentence I used to describe the work of the group—the sentence that starts, "The group's task was to explore and exploit. . ."—that this sentence was a truthful description of the main task of the group that I (and surely most other people in the Meteorological Office who dealt with extending the time frame of the forecasts) understood it to be.*

Dr. Petterssen to Dr. Stagg April 24, 1972

I thank you for your letter of April 17.

I am amazed that I through my earlier letters have not succeeded at accounting for my views. Let me again draw your attention to page 20 of your book where you, explicitly and faultily, describe the circumstances that in 1941 formed the basis for my accepting the invitation to enter into British service. In your above-mentioned letter you write that you did not have any knowledge of these circumstances, and that there has never existed any sensible reason for you knowing this. It therefore seems odd that you are as able to describe these circumstances as you express on page 20.

It seems even more odd that you refer to these circumstances in your information for Dr. Rossby (page 24 in your book). Dr. Rossby was, as you must have known, scientific adviser to the American secretary of defense, who represents high military authorities in the Pentagon. It was not until after the war that I was able to explain to Dr. Rossby the scope of the work I had been responsible for.

For your information, I would like to mention that I was in the Mediterranean area during January of 1944, and that I did not return to England until Dr. Rossby had already departed. After my return, Sir Nelson, his deputy director, and I discussed certain measures which at the time were under consideration in connection with the attack/landing on the European continent. After a long discussion I agreed to continue my work at Dunstable with some changes in the priority of my tasks. I am certain that you will agree with me that your mention on page 20 and 24 in your book must refer to the conditions that existed prior to your informing Dr. Rossby.

I am certain that you will be pleased to hear that I have come to the conclusion that nothing useful may be achieved by continuing this correspondence. As facts are more unflinching than fantasies, unbiased readers of our correspondence should without greater difficulties be able to reach an objective point of view.

*The research I was responsible for was placed under an approved plan. Dr. Stagg's inserted statement therefore seems to be lacking any form of validity. (author's footnote)

Appendix B: Timeline Sverre Petterssen (19 Feb 1898–31 Dec 1974)

1900	Earliest childhood memories in Hadsel [68° 30′ N, 15° 00′ E] on the Eidsfjorden, near Mount Reka.
1906	Family moves to Vik [65° 20′ N, 12° 10′ E].
1911	Family moves to Trondheim.
1913	Leaves school to work in the Trondheim telegraph office.
1915	Enters military officer school with a scholarship that provides *gymnasium* education.
1923–25	As student at Oslo University, attends seminar run by Tor Bergeron, becomes an apprentice of the Bergen school, takes mountain observations with Hans Ahlman (1924), and serves as a forecaster in Oslo (1924–25); B.Sc. (1924).
1925–28	Geophysical Institute, Tromsø, with duties in remote stations, including forecasting for the Amundsen-Ellsworth polar flight of the *Norge* (1926) and rescue operations for the ill-fated crew of the *Italia* (1928); M.Sc. (1926).
1928–39	Bergen, working on methods for computing the movement and rate of development of storms; head of the Bergen regional center (1931–39); Ph.D. (1933); first lecture courses in the United States for the U.S. Navy and Caltech (1935).
1939–42	Massachusetts Institute of Technology, chair of the meteorology department (on leave in 1941–42). *Weather Analysis and Forecasting* (1940); *Introduction to Meteorology* (1941).
1941–45	Meteorological Office, British Air Ministry, adviser on loan from the Norwegian Air Forces; head of the Upper-Air Branch, Dunstable; forecasts for bombing raids over Germany, the Anzio landings, and Operation Overlord.
1945–48	Head of the Norwegian Forecasting Service.
1948–52	Director of scientific services, U.S. Air Force Weather Service.
1952–63	Professor of meteorology, University of Chicago; chair of the department of meteorology (1959–61); chair of the department of geophysical sciences (1961–63).
1963	Retirement in England; emeritus professor of meteorology (1963–74).

Awards

1944	Letter of Commendation from General Dwight Eisenhower. Letter of Commendation from the British Air Council.

1948	Commander, Order of the British Empire.
	Commander, Order of St. Olaf.
	Liberation Cross of Norway.
	Buys Ballot Gold Medal, Netherlands Academy of Science.
1953	Distinguished Service Award, U.S. Air Force.
1958	Gold Award, New York Board of Trade.
	Silver Medal, University of Helsinki.
1962	Charles Franklin Brooks Award, American Meteorological Society.
1965	Gold Medal, World Meteorological Organization.
	Cleveland Abbe Award, American Meteorological Society.
1969	Symons Memorial Gold Medal, Royal Meteorological Society.

Professional Activities

President, Maritime Commission of the IMO, 1939–46.
Chair, Meteorological Division, International Commission on Aerology, 1946–51.
President, American Meteorological Society, 1958–60.
Consultant, President's Science Advisory Committee, 1958–63.
Director, National Academy of Sciences Task Force for ten-year plan for the atmospheric sciences, 1960–61.
U.S. scientific attaché to Scandinavian countries, 1963–65.

Appendix C: Sverre Petterssen, A Select Bibliography

"Kinematical and Dynamical Properties of the Field of Pressure, with Applications to Weather Forecasting." *Geofysiske Publikasjoner* 10, no. 2. Oslo, 1933. 92 pp.

Practical Rules for Prognosticating the Movement and the Development of Pressure Centers. Bergen, 1933. 44 pp.

Værvarsling til Sjøs (with Finn Spinnangr). Bergen: Bergen skipperforening, 1936. 193 pp.

"Contribution to the Theory of Frontogenesis." *Geofysiske Publikasjoner* 11, no. 6. Oslo, 1936. 27pp.

"On the Causes and the Forecasting of the California Fog." *Journal of the Aeronautical Sciences* 3 (1936): 305–09.

Notes on Dr. Sverre Pettersen's [!] Lectures on Synoptic Meteorology. Delivered at the Aerological Officers School, Norfolk, Va. and San Diego, Calif., 1935. U.S. Navy, Bureau of Aeronautics, 1936.

Meteorologi for Sjøfolk (with Finn Spinnangr). Bergen: Grieg, 1938. 160 pp.

"Contribution to the Theory of Convection." *Geofysiske Publikasjoner* 12, no. 9. Oslo, 1939. 23 pp.

"Some Aspects of Formation and Dissipation of Fog." *Geofysiske Publikasjoner* 12, no. 10. Oslo, 1939. 22 pp.

Weather Analysis and Forecasting: A textbook on synoptic meteorology. New York: McGraw-Hill, 1940. 503 pp.

Introduction to Meteorology, 1st ed. New York: McGraw-Hill, 1941. 236 pp.

"Fronts and Frontogenesis in Relation to Vorticity" (with J.M. Austin). *Papers in Physical Oceanography and Meteorology* 7, no. 2. Cambridge, MA: Massachusetts Institute of Technology, 1942. 37 pp.

"Demonstration of Weasel and Other Snow Vehicles, Columbia Icefields, October 15–16, 1942." OSRD–Studebaker. Studebaker Corporation, South Bend, Ind. 35pp.

"Convection in Theory and Practice." *Geofysiske Publikasjoner* 16, no. 10. Oslo, 1946. 44 pp.

Studies in Weather Analysis and Forecasting, 4 vols., edited by Sverre Petterssen. [Final reports under Air Force contracts.] Chicago: Weather Forecasting Research Center, University of Chicago, 1954, 1957, 1960, 1963.

Weather Analysis and Forecasting, 2nd ed., 2 vols. New York: McGraw-Hill, 1956.

Meteorology of the Arctic. Washington, DC: Technical Assistant to Chief of Naval Operations for Polar Projects, 1956.

Cloud and Weather Modification: A group of field experiments. Meteorological Monographs 2, no. 11. Boston: American Meteorological Society, 1957. 111 pp.

Weather Observations, Analysis, and Forecasting. Meteorological Monographs 3, no. 15. Boston: American Meteorological Society, 1957.

Introduction to Meteorology, 2nd ed. New York: McGraw-Hill, 1958. 327 pp.

"Research Aspects of the World Weather Watch." *World Weather Watch Planning Report* no. 5. Geneva: World Meteorological Organization, 1966. 20 pp.

Introduction to Meteorology, 3rd ed. New York: McGraw-Hill, 1969. 333 pp.

Kuling fra Nord: En værvarslers erindringer. Oslo: Aschehoug, 1974. 311 pp.

Weathering the Storm: Sverre Petterssen, the D-Day Forecast, and the Rise of Modern Meteorology. J.R. Fleming, ed. Historical Monograph Series. Boston: American Meteorological Society, 2001.

Biographical Sketches

The Author's and Writer's Who's Who, 2nd ed. Hafner, 1971.

American Men & Women of Science, 12th ed. R.R. Bowker, 1971–1973.

Who's Who in the World, 2nd ed. Marquis Who's Who, 1973.

The International Who's Who, 38th ed. Europa, 1974.

Who's Who in America, 38th ed. Marquis Who's Who, 1974.

The International Who's Who, 39th ed. (Obituary). Europa, 1975.

Who Was Who in America, vol. 7, 1977–1981. Marquis Who's Who, 1981.

Biographical Articles

Karl Johannesen [Obituary], *Bulletin of the American Meteorological Society* 56 (August 1975): 892–94.

R.C. Bundgaard, "Sverre Petterssen, Weather Forecaster," *Bulletin of the American Meteorological Society* 60 (1979): 182–95.

Notes

1. The British author Robert W. Service (1874–1958), bard of the Klondike, went to the Yukon after the gold rush of 1898. His cabin, in Dawson City, is a national historic site. "Robert Service Home Page," http://www.yukon.net/business/Whitehorse/Robert Service1/Home.html (1 December 1999).

2. Oxeye daisy (*Chrysanthemum leucanthemum*).

3. Storms off the North Sea would indeed appear to "come from Mount Reka."

4. In nautical terminology the phrase "all reefs in" refers to the practice of reducing the extent of a sail by taking in or rolling up a part and securing it. *Oxford English Dictionary* (hereafter *OED*).

5. "Strakes" are the continuous lines of planking, of uniform breadth, along the side of a vessel, extending from stem to stern; "clinkering" refers to hardening by fire. (*OED*).

6. I Cor. 15:55, Bible, King James Version.

7. A small variant of the *nordslandbåt*.

8. Quote from John Muir. "The Wild Parks and Forest Reservations of the West," http://www.sierraclub.org/john_muir_exhibit/writings/our_national_parks/chapter_1.html (7 December 1999).

9. Fifteen years later Dr. Robert Millikan, Nobel laureate in physics and a friend of Herbert Hoover, told me a great deal about the behind-the-scenes dealings during the Versailles conferences. In the context of the present account I especially recall his version of an early conversation between the Czech statesman Tomás Masaryk and French President Raymond Poincaré. When the former asked, "What is your impression of Wilson?" the latter replied, "Difficult to place; he talks like Jesus, but he acts like Lloyd George."

10. Named for Tobias Hobson, the English stablemaster; the option of taking the one thing offered or nothing at all *(OED)*.

11. For details see R.M. Friedman, *Appropriating the Weather, Vilhelm Bjerknes and the Construction of a Modern Meteorology* (Ithaca, NY: Cornell University Press, 1989); and Ralph Jewell, "Vilhelm Bjerknes's Duty to Produce Something Clear and Real in Meteorological Science" in G. Good, ed., *The Earth, the Heavens, and the Carnegie Institution of Washington*, *History of Geophysics* 5 (Washington, D.C.: American Geophysical Union, 1994), pp. 37–46.

12. According to Joseph Conrad, art is "a single-minded attempt to render the highest kind of justice to the visible universe"; Henry James maintains "It is art that *makes* life, makes interest, makes importance . . . and I know of no substitute whatever for the force and beauty of its process."

13. It was only years later that I became acquainted with the counterpart, the gentle understatement. For example, "British humour is not to be laughed at."

14. Later, world-famous glaciologist, Swedish ambassador to Norway, etc.

15. Thomas Carlyle, "On Heroes, Hero Worship, and Heroes in History," vol. 3 in *Works*, 18 vols. (London, Chapman and Hall, 1904–06).

16. Aleksandra Mikhailovna Kollontai (1872–1952), Bolshevik, member of Vladimir Lenin's Central Committee, the only woman commissar in Russia's first communist government, and the world's first woman ambassador, served as Soviet minister to Norway from 1923 to 1925. Beatrice Farnsworth, *Aleksandra Kollontai: Socialism, feminism, and the Bolshevik revolution* (Stanford: Stanford Univ. Press, 1980).

17. A gun capable of being discharged on contact, often set as a trap (*OED*).

18. A person having many diverse activities or responsibilities.

19. Roald Amundsen and Lincoln Ellsworth, *Den forste flukt over Polhavet* (Oslo:

Gyldendal Norsk forlag, 1926), transl. *First Crossing of the Polar Sea* (New York: Doran, 1927) paints a generally rosy picture of the expedition, mentioning only the flap over the Italian officers' dress uniforms. The disagreements are recorded in Wilbur Cross, *Ghost Ship of the Pole: The incredible story of the dirigible Italia* (London: Heinemann, 1959).

20. In Finnish a postposition is the opposite of a preposition, appearing at the end of a postpositional phrase. The *Kalevala* (1835), compiled and arranged by Elias Lönnrot, is Finland's national epic.

21. "Arctic Weather of April 15–16, 1928," *Geographical Review* (1928): 556–65.

22. Also called Svernaya Zemlya, Nicholas II Land, or Lenin Land, centered at about 80° N and 100° E.

23. On the crash of the *Italia* and its aftermath see Cross, *Ghost Ship* and Umberto Nobile, *With the "Italia" to the North Pole*, transl. Frank Fleetwood (London: Allen and Unwin, 1930).

24. From 1931 to 1940 Jacob Bjerknes held the professorship in Bergen that had been established originally for his father.

25. Sverre Petterssen, "Kinematical and Dynamical Properties of the Field of Pressure, with Applications to Weather Forecasting," *Geofysiske Publikasjoner*, Oslo, v. 10, no. 2 (1933), 92 pp.

26. King served as fleet admiral and commander in chief of naval forces after the Pearl Harbor disaster in 1941.

27. In 1959, when I visited San Diego on some routine business, the mayor called to present me with the Key to the City, a typical example of American courtesy.

28. *Notes on Dr. Sverre Pettersen's [!] lectures on synoptic meteorology. Delivered at the Aerological Officers School, Norfolk, Va. and San Diego, Calif., 1935*. U.S. Navy, Bureau of Aeronautics, 1936. These notes became the basis for my textbook *Weather Analysis and Forecasting* (New York: McGraw-Hill, 1940).

29. One result of his sojourn at Caltech was Sverre Petterssen, "On the Causes and the Forecasting of the California Fog," *Journal of the Aeronautical Sciences* 3 (1936): 305–09.

30. As a by-product of studies prepared for the Norwegian government, Petterssen wrote the meteorological chapters of P.V.H. Weem's *Air Navigation* (London and New York: McGraw-Hill, 1937).

31. See note 25.

32. Two of the major tenets of National Socialism. The concept of *Lebensraum*, or "living space," was developed by the German geographer and ethnographer Friedrich Ratzel and was misappropriated by the Nazis to justify their military expansion into Poland and other Slavic nations to the east. There, the German master race, or *Herrenvolk*, would rule over a hierarchy of subordinate peoples and organize and exploit them with ruthlessness and efficiency (*Encyclopedia Britannica Online*).

33. ". . . and grow up with the country, " written by John Soule, a journalist in Indiana, and popularized by Horace Greeley.

34. Neville Chamberlain, British prime minister identified with the policy of appeasement toward Hitler; Edward Halifax, British foreign secretary, 1938–1940; and Édouard Daladier, French premier who signed the Munich Pact in 1938.

35. George Rippey Stewart, *Storm* (New York: Random House, 1941); published in several editions, including one for the armed services. The anomaly here is that Petterssen says his conversation about this with Grieg occurred in 1937.

36. Maxim Gorky, V. M. Molotov, K. Y. Voroshilov, and S. M. Budenny were all leading communists.

37. Tyrgve Lie (1896–1968) served as the first secretary general of the United Nations from 1946 to 1952.

38. Major Vidkun Quisling (1887–1945) collaborated with the Germans during the occupation of Norway. The term "quisling" refers to a collaborationist or traitor to one's country. (*OED*).

39. Nordahl Grieg, *Greater Wars* (1940–41), filmscript originally written in English and published in Norwegian as *Større Kriger* (Oslo: Glyndendal, 1990). The script of *Edvard Grieg* is also included in this book.

40. Sigrid Undset, Norwegian novelist and Nobel laureate (1928); Mrs. J. Borden Harriman (née Florence Jaffray Hurst), author, diplomat, and minister to Norway 1937–41; Wendell Wilkie, prominent lawyer and Republican presidential candidate, 1940; Fiorello La Guardia, mayor of New York City.

41. Nordahl Grieg, *Friheten* (Reykjavik: Helgafell, 1943), translated by G. M. Gathorne-Hardy as *All that is mine demand* (London: Hodder and Stoughton, 1944).

42. The Norwegian Air Forces of the army and navy were later consolidated to become the Royal Norwegian Air Force [SP].

43. According to explorations and excavations by Mr. and Mrs. Helge Ingstad, the Norsemen settled on the flats at the northern tip of Newfoundland. Although the climate about year 1000 or earlier might have been better than it is at present, it seems that Leif Eriksson's description of the fertility of Vinland was tinged with propaganda; the same may be true of the descriptions by Leif's father, Erik the Red, of his discovery of what is now known as Greenland.

44. Charles Lindbergh, American aviator who made the first nonstop solo flight across the Atlantic in 1927; Bernt Balchen, born in Norway, was a pilot on Richard E. Byrd's 1927 transatlantic flight and chief pilot on Byrd's flight to the South Pole in 1929.

45. Grieg, *All that is mine demand.*

46. In response to my request eight meteorologists tried to escape. One was killed in a *Luftwaffe* attack on the fishing smack that was to land him in Scotland; one was caught by the Nazis near the Swedish border and suffered only minor punishment, as it was high-level Nazi policy to try to gain the cooperation of Norwegian forecasters; the rest arrived safely via Stockholm.

47. This law, as I understand, states that in the long run everything that can go wrong will go wrong. The identity of the author of this law is not clear, except that he is supposed to be an Irishman.

48. A member of the Women's Royal Navy Service.

49. Meaning it takes every kind of people to make the world go 'round.

50. Berson was a Polish refugee who joined the Dunstable staff soon after the outbreak of the war.

51. Not all stories have a happy ending. Not knowing that Hans was terminally ill, I wrote to him in early May 1972, asking if he, with his enormous card memory, could reproduce the play for me. On his bedside table he left behind a cheerful account of how we had improved our score.

52. Literally "stonemason," typically referring to medieval Italian master artisans working with marble.

53. A play on words in which the noun *Mensch* is used also as a verb. Thus, more or less, "Wherever there are people it will be peopled."

54. I was absent from Dunstable on special assignment and had no part in the preparation of the forecasts of upper-level winds for these raids. Credit for these excellent forecasts must go to the meteorologists at Bomber Command Headquarters and members of my group.

55. V. Bjerknes, J. Bjerknes, H. Solberg, and T. Bergeron, *Physikalische Hydrodynamik, mit Anwendung auf die dynamische Meteorologie* (Berlin: Springer, 1933).

56. In early 1945 the upper-level current of strong winds was "rediscovered" by American pilots flying over Japan (see chapter 23) and soon became known as the jet stream. Since then the literature on jet streams has grown to become one of the bulkiest sections of meteorology.

57. C. H. B. Priestley, Turbulent Transfer in the Lower *Atmosphere* (Chicago: Univ. of Chicago Press, 1959).

58. Spirit of the place.

59. Winston Churchill, *The Second World War*, 6 vols. (London: Cassell, 1948–54), vol. 4, p. 98.

60. In 1939 I had published a paper on the formation and dissipation of various kinds of fog, and I had firsthand knowledge of fog in the Trondheim area. Sverre Petterssen "Some Aspects of Formation and Dissipation of Fog," *Geofysiske Publikasjoner*, Oslo, v. 12, no. 10 (1939), 22 pp.

61. Pierre François Joseph Bosquet (1810–1861), French marshal in the Crimea, who made this famous remark concerning the Charge of the Light Brigade.

62. A series of irregular wavelike ridges formed on a snow surface by wind erosion and deposition, aligned parallel to the direction of the prevailing wind (*OED*).

63. Possibly Cesare Pavese (1908–1950).

64. Although MIT had offered to keep my position open for me, I felt that it was hardly fair to my colleagues and to the department that the steadily growing business should be handled on a temporary basis. In the summer of 1942 I recommended certain changes and promotions within the department and submitted my resignation.

65. *Royal Engineers Journal* 60 (1946): 31–32.

66. Liv lingered on until late April 1945 and died of starvation two weeks before I reached Norway, following the German collapse.

67. Michael Holroyd, *Lytton Strachey: A critical biography* (London: Heinemann, 1967–68), v. 2, p. 80.

68. J. R. Fleming, "Storms, Strikes and Surveillance: The U.S. Army Signal Office, 1861–1891," *Historical Studies in the Physical and Biological Sciences* 30, no. 2 (2000): 315–32.

69. On meteorology in the U.S. military, see Charles C. Bates and John F. Fuller, *America's Weather Warriors, 1814–1985* (College Station: Texas A&M University Press, 1986); and John F. Fuller, *Thor's Legions: Weather support to the U.S. Air Force and Army, 1937–1987* (Boston: American Meteorological Society, 1990). On meteorological education see William A. Koelsch, "From Geo- to Physical Science: Meteorology and the American University 1919–1945," in *Historical Essays on Meteorology*, ed. J. R. Fleming (Boston: American Meteorological Society, 1996), pp. 511–40.

70. "Parlement of Foulys," *The Complete Works of Geoffrey Chaucer*, 6 vols., ed. W.W. Skeat (Oxford: Clarendon Press, 1900).

71. In 1954 Krick paid graceful homage to his sponsors by dedicating his book, *Sun, Sea, and Sky*, to the memory of Dr. Millikan and General Arnold. Irving P. Krick and Roscoe Fleming, *Sun, Sea, and Sky: Weather in our world and in our lives* (Philadelphia: Lippincott, 1954).

72. *Hávamál*, or "Sayings of the High One," is part of the *Poetic Edda*, a thirteenth century manuscript containing mythological and heroic poems composed between AD 800 and 1100.

73. The Boyg (Böygen in Norwegian) symbolizes something that one cannot get through or around; one can neither fight it, nor escape it.

74. Halifax (1881–1959) was an English statesman who served as viceroy to India, (1926–31), foreign secretary (1938–1940), and ambassador to the United States (1940–46).

75. Latin for "first among equals."

76. Hjalmar Riiser-Larsen, *Femtri År For Kongen* (Oslo, Gyldendal, 1957), p. 252.

77. J. M. Stagg, *Forecast for Overlord, June 6, 1944* (London: I. Allan, 1971).

78. Bernard Law Montgomery, *The Memoirs of Field-Marshal the Viscount Montgomery of Alamein* (Cleveland: World, 1958).

79. Patrick Hughes, *A Century of Weather Service: A history of the birth and growth of the National Weather Service, 1870–1970* (New York: Gordon and Breach, 1970), p. 89.

80. The sacred animal of Dionysus was the goat (*tragos*). Referring to the origins of Greek tragedy, Margarete Bieber suggests that "The worshippers of Dionysus danced around the goat, singing the dithyramb; they then sacrificed it, devoured its flesh and

made themselves a dress. . . out of its skin." "Agamemnon Study Guide," http://novaonline.nv.cc.va.us/eli/eng251/agamemguide.html (7 December 1999).

81. Loosely translated, "Oh Lord, save us from experts without knowledge and amateurs without love of the subject"; quoted in Hans Zinsser, *Rats, Lice and History* (Boston: Little, Brown, 1935).

82. Krick and Fleming, *Sun, Sea and Air*.

83. Hughes, *A Century of Weather Service*.

84. Theodore von Kármán, with Lee Edson, *The Wind and Beyond: Theodore von Kármán, pioneer in aviation and pathfinder in space* (Boston: Little, Brown, 1967).

85. Josephus attributes a similar saying to Zorobabel, governor of the Jews under King Darius: ". . . we also ought to esteem truth to be the strongest of all things." *The Works of Flavius Josephus*, transl. William Whiston (Philadelphia: Lippincott, Grambo, 1850), Book XI.

86. Appleton was awarded a Nobel prize in 1947 for his discoveries in ionospheric physics.

87. Devik, a Norwegian physicist, was then undersecretary of education.

88. Marshal Philippe Pétain, head of Vichy France, 1940–44.

89. Literally, "somewhat later future."

90. A light and small two-wheeled carriage or cart used in India (*OED*).

91. A man of much or varied learning; a great scholar (*OED*).

92. Weather warfare was indeed practiced in Indochina in the 1950s and 1960s.

93. At the times here considered H.R. Haldeman and John Erlichman were, respectively, White House chief of staff and chief adviser on domestic affairs to President Nixon.

94. Richmond Park is a nature preserve southwest of London. http://www.guidetorichmond.co.uk/park.html (7 December 1999).

95. Harriet Augusta Dorothea Löwenhjelm (1887–1918), Swedish poet, painter, and author.

INDEX